PERSPECTIVES ON THE ENV

Perspectives on the Environment 2

Interdisciplinary research on politics, planning, society and the environment

INTERDISCIPLINARY RESEARCH NETWORK ON THE
ENVIRONMENT AND SOCIETY

Edited by
SUE ELWORTHY
KEVIN ANDERSON
IAN COATES
PIERS STEPHENS
MATT STROH

Avebury Studies
in
Green Research

Routledge
Taylor & Francis Group

LONDON AND NEW YORK

First published 1995 by Ashgate Publishing

Reissued 2018 by Routledge
2 Park Square, Milton Park, Abingdon, Oxon, OX14 4RN
711 Third Avenue, New York, NY 10017

Routledge is an imprint of the Taylor & Francis Group, an informa business

Publisher's Note
The publisher has gone to great lengths to ensure the quality of this reprint but points out that some imperfections in the original copies may be apparent.

Disclaimer
The publisher has made every effort to trace copyright holders and welcomes correspondence from those they have been unable to contact.

A Library of Congress record exists under LC control number: 94074547

ISBN 13: 978-1-138-32159-5 (hbk)
ISBN 13: 978-1-138-32163-2 (pbk)
ISBN 13: 978-0-429-45247-5 (ebk)

Contents

Contributors

Margaret Anderson is a senior lecturer in the Department of Environment, Wye College, University of London.

Mark Baker is a PhD student at Aston Business School, Aston University.

John Barry is a lecturer in the Politics Department of the University of Keele.

Raymond Bryant is a lecturer in the Department of Political Studies, School of Oriental and African Studies, University of London.

Ian Coates is a PhD student in the Sociology Department, University of Bristol.

Ute Collier is a lecturer in the Division of Environmental Sciences, University of Hertfordshire.

Bob Evans is a senior lecturer in the School of Land Management and Urban Policy, South Bank University.

Åse Berit Grødeland is a PhD student at the Institute of Soviet and East European Studies, University of Glasgow.

Michael Jacobs is a research fellow at the Centre of Environmental Change, Lancaster University

Les Levidow is a research fellow in the Centre for Technology Strategy, the Open University.

James Meadowcroft is a lecturer in the Department of Politics, University of Sheffield.

Julia Meaton is a lecturer in the Department of Geographical and Environmenal Sciences, University of Huddersfield.

Ali Ghanbari Parsa is the Aubrey Orchard-Lisle research fellow at the School of Land Management and Urban Policy, South Bank University.

William Penrice is Coordinator for International Relations, Wakayama City, Japan.

Simon Shackley is a research associate at the Centre for the Study of Environmental Change, Lancaster University.

Elizabeth Shove is a senior lecturer in the Buildings and Society Research Unit, Department of Social and International Studies, University of Sunderland.

Acknowledgements

Putting together a book such as this one requires the high level of cooperation and sharing of ideas that IRNES exists to foster. First we must thank Avebury for their continued support and encouragement. Sarah Markham, Avebury's editor of this series, was friendly, efficient and believed that, despite all odds, the editorial team would meet the agreed deadline with an interesting book. We thank Michael Jacobs for writing the foreword, Lesley Morris for coordinating the gathering of disks from the contributors, Cheryl Benton for typesetting and, of course, the authors who not only adapted their papers to fit the scheme of the book in terms of intellectual coherence but also performed skilled wonders on their computers adjusting to the agreed format.

We thank the ESRC for the financial support given to IRNES, and GEC for its most welcome contribution to the conference finances. Warwick Research Institute undertook the typesetting of this volume, for which we are most grateful.

Declaration

The views expressed within the individual papers of this book are those of the respective authors and do not necessarily reflect either the views of IRNES or of the ESRC.

Foreword

Michael Jacobs

"The environment", Albert Einstein remarked, "is everything that is not me". This doesn't narrow the research field down very far. Fortunately, there are ways of dividing the subject into more manageable categories. In practice we may observe three types of environmental research, corresponding to three different conceptions of the environmental subject.

The first, and oldest, belongs in what we might call the natural science paradigm. As recently as thirty years ago, scientific research was practically the only kind of research there was on the environment. Originally, indeed, there was no environmental "subject" as such at all - simply aspects of biology, geology, physical geography and so on. More recently, new holistic sciences have emerged - ecology, oceanography, climatology; many universities offer more or less integrated courses in "environmental sciences". Nevertheless all these disciplines essentially treat the environment in the same way. It is a given physical thing, made up of living and non-living objects which exist in nature and whose interactions - with each other and with humankind - can be observed by the application of a positive scientific method. Increasingly, such research has led scientists to advocate specific environmental policies, drawing on more or less explicit assumptions about the desirability of different "natural" states.

In the last twenty years, and especially in the last five or six, economics has entered the environmental field, the first and much the most dominant of the social sciences. The economics paradigm also sees the environment as a physical thing, which can be observed through scientific methods. Indeed, it starts - more or less unquestioningly - from the information furnished by the physical sciences. But for economists the emphasis is on the environment's *economic value*. The environment, they argue, provides a range of goods and services to the economy: the provision of resources

1

and energy, the assimilation of wastes, and various "amenities" and "life support mechanisms" such as climate regulation and the maintenance of genetic diversity. The purpose of environmental economics is to understand the ways in which these goods and services are *valorised* - and, where they are not, to reveal or impute appropriate economic values. By this means, economists are able to understand environmental objects and activities as economic commodities, and can locate them in the same framework as the ordinary "produced" commodities of standard economic analysis. Again, given various more or less stated assumptions about economic and political goals, conclusions for environmental policy then follow. The advantage of this approach, the economist argues, is that it understands not simply physical environmental change but its "costs" and "benefits" as well.

The non-economic social sciences and the humanities have been relatively late arrivals on the environmental scene. Yet in the last few years there has been an explosion of research activity, in - and across - disciplines as diverse as sociology, politics, anthropology, human geography, social psychology, philosophy, religious studies and cultural analysis. For students in these fields, it is not the physical nature or the economic value of the environment which is of interest, so much as its *meaning*. For the social sciences and humanities, the environment is not a set of things, but a site and a symbol of human identities and social relations.

And this leads to a quite different understanding of the environmental subject. Indeed the environment as a physical and economic entity - in the reified form it takes within the natural sciences and economics - dissolves under the gaze of the new research. Sociologists of science, for example, ask exactly how scientists make their claims on truth: how the sociological and political practice of scientific activity contributes to its conclusions; how scientists handle the endemic and critical problems not just of risk but of uncertainty and ignorance. Philosophers question the economic account of value, exploring the diverse (and not all human-centred) ways in which the environment can be said to be valuable. Anthropologists and cultural analysts identify the symbolic roles played by "the natural world" in different societies and social groups. In late modern industrialized societies such as Britain, for example, environmental concern is increasingly being understood as symbolic of a wider disquiet at the trajectory of technological development and the fragmentation of traditional identities and communities - a process in which actual physical change can become of almost secondary significance.

The striking thing here is not simply that "the environment" becomes under this perspective a socio-cultural rather than a physical phenomenon. It is that the research becomes self-reflexive. It is concerned with understanding how the environment is understood - not least, by the dominant scientific and economic paradigms.

A feature of the new research is the way in which the environment becomes an arena in which other issues - often ones of long pedigree - can be played out. The widely-discussed concept of "sustainability", for example, is commonly presented in terms of justice or equity, both between rich and poor nations and between current and future generations. In this it draws on and adds new perspectives to questions long debated in political philosophy and economy. Environmental policy raises critical questions about the governability of modern societies, in a world of global pollution and global information and capital transfer: both notions of democracy and issues of economic power are challenged afresh. Environmental awareness has revived interest in cosmology and theology - the study of humankind's place, and its understanding of that place, in the "natural order".

The implications of the new research for environmental policy making are large. The new perspectives challenge the positivism of the scientific and economic paradigms, questioning their understanding both of physical environmental change and of society - indeed, refusing to separate the two. But - contrary to the claims sometimes made from within those disciplines - they do not paralyse policy making. They inform it. The new perspectives open up policy debate to more critical viewpoints. They reveal the social assumptions and political interests tied to particular policy recommendations. They demonstrate the power that environmental meanings have to influence individual behaviour and social institutions - not least, to affect the acceptability and purchase of environmental policy itself.

Like its predecessor volume, *Perspectives on the Environment 2* is a product of the new research. The papers in it illustrate the wide variety of understandings brought to the environmental subject, not just by new disciplines, but by the inter-disciplinarity which is their hallmark. In different fields they show how these understandings can illuminate both what is happening in the world and how, through carefully designed policy, that can be changed. The environment that emerges from this book is physical and economic, yes, but it is also social, cultural and political. It may be true that the environment is not about "me". But as this book shows, it is quintessentially about *us*.

Professor Michael Jacobs
Centre for the Study of Environmental Change
Lancaster University, 1994

Introduction

The essays in this book have been selected from papers presented to the second conference of the Interdisciplinary Research Network on the Environment and Society (IRNES), held in Sheffield in September 1993. IRNES was established in 1990 on the initiative of a group of postgraduate research students, subsequently gaining recognition and financial support from the ESRC through both its seminar programme and its Global Environmental Change initiative. The second IRNES conference built on the success of its first conference held in Leeds in 1992, which resulted in the publication of the book, *Perspectives on the Environment: Research and Action in the 1990s.*[1] This was produced to disseminate the work of IRNES members to a wider audience, a process which this second volume seeks to continue.

The book is divided into four sections. Part I explores the social dimensions of environmental technology beginning with Simon Shackley who examines how global climate models are constructed and used by scientists in order to influence policy decisions. The author uses insights drawn from the sociology of science to question whether such models provide the best available means of predicting the possible consequences of global warming. Les Levidow provides a critical analysis of discourses associated with the application of agricultural biotechnologies. He argues that the military metaphors used by the biotechnology industry are indicative of its manipulation and domination of the environment. Elizabeth Shove considers how the buildings we inhabit create an artificial climate to protect us from the natural elements. She explores the social practices which different indoor environments and building technologies encourage, and the environmental implications of this in terms of the consumption of energy and resources.

Part II examines national and international politics of the environment. In his historical investigation of environmental concern in Britain between 1919 and 1949, Ian Coates shows how this took a variety of forms, involving groups from all sectors of society. He also draws parallels between the diversity of environmental concern in this earlier period and some contemporary forms of environmentalism. Åse Grødeland focuses on the Ukrainian Greens and seeks to evaluate the charge that the movement principally represents nationalist and separatist tendencies. After considering the influence of culture and religion on environmental politics, Grødeland argues that the incorporation of national sentiments by the Ukrainian Greens has not taken a negative or parochial form. Rather this can be seen as part of an attempt to reassert local democratic control over an environment seriously damaged by the central command economy of the Soviet Union. Raymond Bryant's examination of forestry management regimes in Burma, Thailand and Indonesia explores the relationship between colonial and postcolonial practices. He shows that the policies of these countries need to be understood in terms of political and economic factors which affect the control of forest resources at local, national and international levels.

Part III deals with planning for sustainability, concentrating on land use planning and continues the national and international theme developed in Part II. Ali Parsa and William Penrice examine the Japanese planning and development process to determine whether there is room for environmental considerations in decision making at local and national levels. Bob Evans draws attention to the fact that the British planning system was established with the assumption that there would be public ownership of development value. Though this dimension of the planning system was soon abandoned, he argues that its reintroduction is essential if sustainable development is to become a reality. Mark Baker uses a case study in the West Midlands to demonstrate the institutional and policy weaknesses of the British planning system in attempting to apply the concept of sustainability in practice. On a more optimistic note, Julia Meaton and Margaret Anderson develop a methodology to show that it is possible to involve the public actively in the planning process.

Part IV considers the role of the state regarding the issue of environmental protection. John Barry argues that, rather than being seen as a modern variant of anarchism, green ideology can encompass a theory of the state. Since Barry sees the democratization of the state as integral to green politics, it follows that green political thought can be understood as operating critically within the framework of modern liberal democracy, instead of being regarded as a

6

fringe ideology standing outside it. The democratic theme is taken further by James Meadowcroft, who examines various explanatory theories of democratic politics to explore how political institutions could address environmental issues more adequately. Finally, this problem of state policy and the downgrading of environmental issues is explored by Ute Collier who investigates how it relates to participation in the European Union. Focusing on the principle of subsidiarity, she wonders whether this will be used by member states as a means to avoid implementing measures intended to tackle global warming.

Thus the book ends as it began with the subject of global warming. This issue particularly reveals the dilemma common to many environmental problems, in that though the scientific understanding of the processes involved may be contested or incomplete, the consequences of delaying action until the evidence is more conclusive could be potentially disastrous. Whilst there are indications that significant climatic changes are taking place, it could also be argued that policies to reduce human impacts on the environment are justifiable ends in themselves and will have favourable social and economic consequences.

As these papers make clear, the environment is not just a scientific issue, but also requires the consideration of social, political and economic structures, and of the values, perceptions and practices which determine human interactions with the natural world. The inclusion of these dimensions highlights the need for interdisciplinary research, and it is for this reason that IRNES aims to bring together researchers from all disciplines with an interest in studying the environment. By stimulating debate, sharing information and encouraging cooperation, IRNES hopes to contribute to the increased understanding of these issues which is required for the formulation of long term solutions to local, regional and global environmental problems.

Notes

1. IRNES (1993), *Perspectives on the Environment: Research and Action in the 1990s*, eds Holder,J., Lane P., Eden,S., Reeve,R., Collier,U. and Anderson,K., Avebury, Aldershot.

Part I
SOCIAL DIMENSIONS OF
ENVIRONMENTAL TECHNOLOGY

1 Mission to model earth

Simon Shackley

In this chapter the relationship between science and global environmental policy is explored using climate change as a case study. I will present four different accounts of the interpolation of science and policy: scientism; science as politics by other means; scepticism; and new forms of science.[1] Some of the examples used are very specific and microsociological, reflecting the use of ethnographic or close observational methods; other examples draw on existing scholarship in the social sciences. The aim is to assess critically what different approaches might offer in the way of useful insights into the relations between knowledge, policy and action, and to briefly raise some normative issues which arise.

Scientism

According to many, science is the most authoritative source of knowledge of the environment and good policy must, therefore, be based on good science. This linear extension from science to policy is a form of scientism, the extrapolation of scientific idioms and methods to conventionally or currently nonscientific realms in the belief that the conditions which make science authoritative and effective in producing knowledge of the natural world are applicable in those other fields and should be applied.[2] An example is provided by the evidence of the former head of the UK Meteorological Office, Sir John Mason, before a Select Committee of the House of Lords. He commented that:

it is very important in my view that any major policy that you make, which is likely to have either major economic or social consequences, should as far as possible be based on good science. If it does not make good scientific sense, it almost certainly will not make good economic sense and in the long run will not make good political sense. If it is basically a scientific and technical problem not based on good science and good technology and real understanding of that, then I think you are in a dangerous situation.[3]

According to this view, science, as a unique source of knowledge of the "real world", circumscribes what is real and hence feasible and desirable in economic and political terms. The natural and social worlds are therefore seen as continuous realities, in which failure to recognize actual limitations (which are naturally given, whether by physical or socio-economic laws) has serious consequences for policy success.

A further example of scientism is provided by elements of the UN sponsored Intergovernmental Panel on Climate Change (IPCC). The IPCC has three working groups dealing with: scientific assessment (WGI), impacts (WGII) and "cross cutting issues", including emission scenarios, economic and social assessments (WGIII). The IPCC, especially in its early phase form 1988 to the early 1990s, appears to have adopted a prescriptive linear model of policy, illustrated in figure one, in which the natural sciences drive the assessment. The WGI report was widely ascribed more credibility in policy and scientific circles than that of WGII and, especially, of WGIII, and the natural sciences bias is indicated by the leading role of environmental scientists in the current WGs II and III.[4] The Chairman of WGI has commented that: "We have been given the Earth to look after, and we have also been given the science and technology to do it":[5] a notion of technocratic stewardship which frequently accompanies scientism.

Figure One: The linear model of environmental science and policy

Natural Sciences   Social Sciences   Politics

more authoritative & certain   messy & irrational

At a meeting of the IPCC in December 1993 one of the issues discussed was a forthcoming workshop between all three Working Groups to help interpret the Climate Convention signed at Rio in 1992, in particular the meaning of the aim of preventing "dangerous anthropogenic interference with the climate system".[6] An early draft programme for the workshop, drawn up by WGI, adopted the linear model encapsulated in figure one. Hence the sequence of proposed sessions went from talks about emissions and atmospheric concentrations of greenhouse gases, to ones on climate simulations, ecosystems, socioeconomic systems and, finally, risks and integrated models, with only about a fifth of the papers being social scientific. The impacts of climate change were defined almost exclusively in terms of effects upon discreet ecosystems or managed primary resource systems.

Participants from WGs II and III argued that this sequence was inappropriate since the key issue was the sensitivity of different systems to climate change; once this was established a range of simulations could be applied to help define what level of change might be considered to be dangerous. One WGI participant objected to this on the grounds that it went from less certainty to more certainty, his assumption being that the usefulness of knowledge is defined more by its certainty than its policy relevance. The agenda was changed, however, to consider sensitivity at the outset of the meeting.[7] Yet, although a number of socioeconomic systems, including health, energy and fisheries, were considered during discussion of the agenda, it was decided to restrict the initial survey of sensitivity to natural ecosystems. It was felt to be too ambitious to look at socioeconomic systems such as energy, presumably because their uncertainties were thought to be too massive. Although there were several strands of argument influencing agenda setting here, scientistic assumptions were highly visible, including conceptions of scientific certainty and useful knowledge and the discretization, on the model of the natural sciences, of the social into socioeconomic systems with differential sensitivities.

Criticisms of scientism

The criticisms of scientism are wide ranging and well known.[8] One common argument has been that the social realm has quite different dynamics and epistemology from the natural. For example, if social reality is socially constructed then culture, institutions, political systems and understandings will influence, or, some would say, wholly determine, how the environment is viewed and treated by individuals and organizations. However natural scientists define "dangerous climate intervention", therefore, this may or may not bear comparison to "lay" understandings. Seen from the social end, the scientific definition may be readily rejected if it does not correspond to the definition which is achieved through political negotiation; there is no reason why science, even if it is "right", should prevail.

An example of disparity between scientific and lay formulations is provided by Le Roy Ladurie[9] who notes that the southern limit of olive growing in France moved steadily northwards during the Little Ice Age, and especially during the cold decades of 1550 to 1600. "Paradoxically, the northern limit of the olive in France did not move southward until the twentieth century, and that right in a warming up period in which in theory the opposite might have been expected".[10] Le Roy Ladurie relates these migrations to newly emerging market opportunities in the 16th century and to new sources of competition in the 20th century; in other examples religious beliefs and institutional commitments have been shown to be a key influence.[11]

The conflation of science and technology has been a particularly powerful expression of scientism, at least in policy cultures in the UK. If technological change, say to limit greenhouse gas emissions or alternative energy production systems, is beholden to precise scientific knowledge inputs, such as risk assessment and corresponding amelioration and mitigation strategies, some evidence suggests that the independent process of technological innovation can be inhibited. If so, this not only limits the availability of technological options but also curtails economic performance irrespective of the success of the scientific programme.[12]

It is of course true that without science many global environmental change issues would not have been defined and articulated in the same way, or even at all. It is difficult to imagine, for example, that the ozone hole, or a global mean warming of 0.5oC in the last century, could have been known without the methods, practices and infrastructure of science. Yet the scientific definition of an issue is far from constituting a global environmental problem. There was a fairly strong scientific consensus on the likely effects of greenhouse gas emissions from the early 1980s onwards, yet the problem of global warming only surfaced on the wider policy agenda in the wake of the North American drought of 1988.[13] Some have argued that there is a "developed world bias" in the attention given to global warming; according to Parikh, for example, there are more important priorities for the south such as unsafe drinking water, lack of sanitation facilities and unclean fuels.[14]

It is also clear that science cannot provide the synoptic overview of an environmental issue which is sometimes assumed from the perception of a few areas of apparent

14

certainty[15] let alone prescribe and institute credible and effective policy responses and social actions. So, for example, a record of half a degree celsius warming in the past century cannot be attributed unambiguously to anthropogenic emissions of greenhouse gases since it may be natural variability. The scientific basis for concern about global warming is more rooted in the findings of general circulation models (GCMs) of the climate system. Yet, there is a basic indeterminacy in the scenarios of climate change from such models since there is no way of confirming their predictive success in advance of the events being predicted.[16] Whilst the GCM's successful simulation of the current climate may be cause for some confidence, the longer term feedbacks which will influence the response of the climate system on the decadal and century timescales, cannot automatically be confirmed in this way.[17] There may, for example, be important biological and chemical feedbacks which are not currently included in GCMs. Modellers themselves point to the low credibility of the regional scenarios of climate change from GCMs;[18] hence it is at present difficult to envisage a robust science-led policy extending beyond global greenhouse gas emission reductions in response to a globally defined threat. However, scholarship on international environmental negotiations reveals the degree to which they are steeped in political and economic arguments and their contingent connection to other strategies and agendas.[19] It is, indeed, hard to imagine such processes being "driven" by science.

The rhetorical force of scientism

The above comments are critical of scientism on two counts. There is the point that the advocates of scepticism rarely make explicit or justify their assumptions, for example that the natural and social worlds are continuous. They are being selective in which strands of their argument they wish to raise and argue for, as if beyond that we should simply trust in their judgements. The second point relates to the implicit values behind scientism, such as the a priori value ascribed to scientific and technical information. These criticisms do not, however, deny the rhetorical force of scientific arguments. That force can be glimpsed in the disparity between the supposedly dispassionate and neutral methods of science and the impassioned, fervent pleas, apparently based on science, to secure the future of the planet.[20] For example, Sir Crispin Tickell, a diplomat who played a key role in putting the climate change issue on to the UK and international agendas, recently commented that science is the "best weapon" with which to persuade politicians of the importance of global environmental issues.[21] A former chief scientific advisor to the British government recalled how climate scientists had

15

shown senior politicians predictions of climate warming stretched out over the next hundred years that he thought were intended to startle them into policy action.[22] An example of one such projection, which was shown in 1990 to the British Prime Minister, Mrs Thatcher, during a presentation of the major findings of the IPCC WGI, is illustrated in figure two.

Figure Two: Simulation and prediction of global mean temperature resulting from business-as-usual emissions

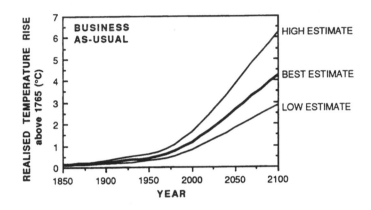

Source: IPCC (1990) figure 8, p.xxii.

How this figure was presented at that meeting is, unfortunately, not known. The rhetorical force of such graphs is their claimed predictive power and immutability in a wide range of different contexts, from ministerial offices to industry boardrooms. The uncertainties and indeterminacies of the knowledge behind them are black boxed;[23] in this instance critical assumptions regarding population growth, economic activity, energy intensity and land use change are deployed to generate a set of emission levels of CO_2 and other greenhouse gases which together construct the scenario of "business as usual". Such projections effectively reify an enormous range of social processes into a metric amenable to scientific enquiry. An even stronger scientism creeps in when the range of scenarios constructed is taken to represent the credible range of societal adaptability and response, and as such is used to define the "realistic" set of policy alternatives, or is used as an input to further research.

Science as politics by other means

A very different approach from scientism is to regard science as irredeamiably imbued with political, institutional or structural interests; or, after Latour, "science is politics by other means". The scientific community, or sections of it, can itself be regarded as a political agent, attempting to secure resources for its own survival and expansion. Boehmer-Christiansen claims, for example, that a scientific policy elite at the IPCC WGI manicured the representation of the scientific uncertainties in its reports, and through other avenues, such as meetings with high ranking policy officials, so as to put the issue of climate change on the political agenda as well as making itself indispensable as a source of expertise. Implicit in this account is the notion of elites as essentially self serving and interested in securing autonomy, resources and prestige. Scientific elites have generally been excused the attribution of less than respectable motives because the objectivity of "the scientific method" and the knowledge so generated, was thought to maintain its integrity beyond reproach.

A problem with such political accounts, however, is their reliance on an a priori and stable definition of interests since on what basis can we attribute interests to actors? The problem is clear when the actors are large, multifaceted institutions or heterogenous collections of individuals and networks, since multiple, perhaps contradictory, articulations of "interest" are routine, but qualitative research has also indicated widespread ambivalence at the level of the individual actor.[24] A "single story" is therefore unlikely to account for different understandings of what autonomy, resources, prestige or the furtherance of scientific knowledge mean, in practice, to different people and organizations.

Complexity argues, therefore, against a simple or single resolution of the relation of science and politics. Some cases do, however, appear somewhat clearer cut, such as the so called "greenhouse contrarians", a small group of vociferous scientific sceptics of global warming from the USA. These meteorologists and applied climatologists are well known for their attacks on the IPCC consensus, and many of their questions and arguments are fair challenge to the mainstream position raising, for example, some of the above mentioned indeterminacies and assumptions. Yet it is difficult to avoid concluding that a major motivation behind the contrarians' posture is a commitment to strongly held values and to social, economic and political practices for which the USA is famous. There are, for example, clear inconsistencies and misleading representations of certain scientific issues in some of their discussions.[25] One example I observed at a meeting was when a contrarian compared the output from a transient GCM run, in which atmospheric CO_2 concentration is increased yearly by a standard increment for 70 or 80 years, to the history of climate change since 1900. The problem with such a comparison is that no modellers have ever claimed that the transient runs are attempts to simulate the climate of the 20th century and the absence of aerosols and greenhouse gases except for CO_2 in the model, would seem to indicate this quite clearly.[26]

In my interviews and informal conversations with some of the contrarians, I have gained the distinct impression that a powerful belief in individualism, and distrust of and rejection of control through government institutions and other bureaucracies, is a major influence upon them.[27] The contrarians put great confidence in human adaptability to what ever conditions turn out to be pressing and in the potential for technological innovation. To individualists, global warming looks like the ideal means by which bureaucrats can attempt to increase centralized control, since the policy measures span a huge range of domains and activities. This example suggests that it is sometimes possible to identify the influence of political values on how scientists argue in a controversy. The proviso is to distinguish that from a different, more widespread, level of scepticism amongst scientists, in which some of the specific and more general framing assumptions, concepts and practices behind knowledge of climate change are up for debate, and which cannot be characterized in the same way as for the contrarians.

Scepticism

A programmatic, and sometimes radical, scepticism towards scientific knowledge is one approach within the sociology of science. The claim that scientific knowledge is in some senses superior knowledge, hence uniquely rational, objective or universal, comes under critical scrutiny, by inter alia detailed, micro level studies of scientific reasoning and practice in the work-place.[28] How then should closure around particular knowledge be understood? There are different responses in sociology of science, from exploring social processes, institutions and commitments, to analyzing how networks of scientists, institutions and nonhuman actors, such as greenhouse gases or the algae which respond to them, become stable and durable. One of the key theoretical issues of disagreement is the degree to which closure can be reduced to social processes, especially if the latter are themselves undergoing change.[29] These sorts of approaches are especially attuned to the ways in which changing knowledge claims and new social processes, forms of organization and alliance become mutually producing and justifying.

An example is to explore how GCMs have come to play a major organizing, synthesizing and legitimating role in climate change research. Rather than just accepting the argument that the role of GCMs can be understood as a function of those models' status as the best, most sophisticated science, the challenge is to describe the holistic sets of relations which together ascribe this status to GCMs. Three overlapping sets of reinforcing relations can be identified.[30] Firstly, there are the connections between climate modelling science and policy making: GCMs are unique amongst climate models with regard to their potential for producing detailed regional scenarios of climate change, though they are currently far from achieving this with any credibility. That predictive capacity would, in principle, permit detailed policy measures to be taken at the regional level across a whole range of policy domains, including water resources, coastal management, agriculture, transport, industrial development, health and so on. This micro level environmental management could, in principle, be applied globally, although with local specificity. A mutual reinforcement between the modellers and the existing and emerging communities engaged in environmental and resource management, planning and control may come to support the hegemonic position of GCMs.

Secondly, there are those networks between GCM modellers and other scientists from a range of disciplines which are presently engaged in the further development and improvement of GCMs; by, for example, inclusion of atmospheric chemistry, and biological processes. GCMs are sufficiently complex that representation of such new processes can occur at a reasonably high level of sophistication, compared with simple models in which a very basic mathematical representation might be used; hence those specialists have less to sacrifice in terms of reducing their complex knowledge to a few formalisms when it comes to GCMs compared to simple models. There are, therefore, mutually sustaining overlaps between GCM modellers and extradisciplinary specialists.

If we "backcast" the process outlined above, it could be argued that development of GCMs has arisen through a compromise between maintaining the complexity of particular specialist knowledge on the one hand, and simplifying and standardizing knowledge sufficiently that it can become combined and aggregated in GCM building and development on the other. Hence the position of GCMs as "best climate models" has perhaps emerged through the continual calibration of the appropriate model design and level of complexity between those with responsibility for model synthesis and those whose specialism is being incorporated,

19

whether it be cloud physics, radiation physics or oceanography. The synthesizers benefit from the credibility given to their reductionisms by the participating experts, whilst the specialists gain the opportunity to study how detailed processes interact with other components of the climate system. My third argument is that the perception of GCMs as the "best models" arises through a set of relations and negotiations between "synthesizers" and "specialists" which is ever changing, such that in the course of time the boundary between outside expert and modeller becomes blurred. The legacy of such past judgements, together with the aforementioned connections to new outsider scientists and policy makers, helps to create a stable actor network, in which GCMs are hegemonic. More controversial is the role of "nonhuman actors", such as the global trend of increasing atmospheric concentration of CO_2 and the century long record of global mean surface temperature. Such apparently certain "facts" may be important in stabilizing relationships between scientists, policy makers, their institutions, greenhouse gases, the global circulation and measurement technologies.

Criticisms of scepticism

One possible problem with programmatic scepticism is that it undermines the authority of science and hence the seriousness and legitimacy of global environmental phenomena. This sentiment is expressed well by Geoff Mulgan:

> .... the green movement has itself contributed to [the] decay of deference and authority. It was nurtured in a libertarian, anti-statist culture that doubted government scientists and ministers, and was prepared to show why their authority was fatuous. Militant scepticism was a good tool for blocking grandiose developments. But now that the greens need a legitimate authority to take tough decisions in the name of the future, they find it is on longer there.[31]

It is critical, however, to reject this argument in so far as it refers to the rhetorical force of scientific argumentation; unless, that is, one accepts this form of rhetoric as positive. My reasons for not doing so are that such forms of persuasion seem to generate a "politics of fear", in which the rationale and motivation for change is some dreadful consequence or global catastrophe, knowledge about which has to be taken almost entirely on trust. I find it difficult to believe that the politics of fear is the right or desirable way to create the conditions for change.[32] Scientific knowledge is well suited to function as a cajoler because it is highly specialized, inaccessible and frequently imbued with exceptional properties. Yet when used as such the indeterminacies and ignorance of much scientific knowledge are all too readily glossed over, and the prospects for future information and understanding frequently embellished. Very often it is those who represent science to policy makers, the "gate keepers", who assume responsibility for judging how to represent scientific uncertainties and future knowledge prospects, rather than the practitioners. It is not necessary to suggest that there is a single correct way of representing a body of scientific knowledge to argue that some representations by gate keepers are disingenuous. This is not an argument against using expert knowledge in decision making, therefore, but one in favour of a more open representation of scientific knowledge which acknowledges its variegated character. In answer to Mulgan, "legitimate authority" does not flow automatically from science, but has to be earned by the institutional cultures of science, for example through their willingness to trust in their publics' ability to cope with scientific indeterminacies.

Fear of global collapse may provide a powerful rationale for change, but frequently sidelines debates over what kinds of societal and epistemic changes might be desirable. A further problem with scientism is that it reduces the multiple sources of environmental concern to scientific knowledge; hence it ignores the potential role of these other forms of understanding in catalyzing change. Similarly, there are multiple rationales for policies such as energy efficiency, emission reductions, and better public transport; to argue for them on the basis of scientific knowledge alone might fail to capture the level of support they deserve (Wynne 1994). Indeed, a sociology of science position might well regard the presumed superiority of science to other types of knowledge and practice as a potentially major hurdle to creative and effective policy responses and societal change for example through supporting an implicit scientism.

New forms of science

My last category, new forms of science, adopts a posture of constrained scepticism. In this approach it is argued that the present scientific endeavour is in some senses flawed, for example through adoption of reductionist methodologies, overly mechanistic modes of analysis, overspecialization and segmentation of research activity, overbureaucratization of research organization, and so on. The implication is that if scientific thought and practice were to change appropriately, the credibility, and hence authority, of scientific knowledge would be enhanced.

There are many examples of new forms of science which could be included here, such as philosophical critiques of reductionism, Gaia theory, and new ideas on chaos, complexity and simplicity. Here, however, I will briefly consider the ideas of Henk Tennekes, who has a background in GCM modelling, but now considers there to be basic limitations arising from the conceptual assumptions behind climate modelling.[33] Tennekes argues against the technicalization of the issue: "the climate problem is a problem of human behaviour; the whole machinery of the physical sciences is almost entirely irrelevant".[34] Alongside this is a much more humble analysis of the current and likely future achievements of climate science. Whereas in the opinion of the mainstream modellers, the current limitations of GCMs are not so huge as to invalidate their output, so in some senses justifying the further development of the models, Tennekes is less convinced of the models' present worth: "current models are *grossly* inadequate". He points out that the simulation of current climate is enabled by the use of "heavy handed" model tuning and large ad hoc corrections, argues that there is no evidence of the predictive skill of climate GCMs and even doubts whether climate will be any more predictable than weather. As for model development, he notes that: "Current models represent say, 0.1% of all processes and feedback loops on the planet, and models 10 times as sophisticated still imitate only 1% of reality. For the time being, research will *add* to the complexity of the problem, and I doubt that we'll *ever* be able to make *any* quantitative prediction". Tennekes is not just sceptical of GCMs, however, but of environmental science more generally, and proposes a search for what he terms a "new ecological grammar". He argues for a qualitative approach in which understanding, values and humility are central. Quantitative approaches, in which notions of "solving the problem" are central, are antithetical to this new approach.

One problem with some such accounts, though not that of Tennekes, is that they are often overly cognitive, that is they concentrate on the intellectual changes needed as if these were divorced from the social and institutional contexts of research. Such critiques therefore neglect the key point that the character of scientific knowledge is socially and institutionally shaped. An analysis which fails to account for this is missing out a vital influence on the establishment of particular knowledge claims and approaches as credible and superior. A further consequence is that the "needs" of policy making rarely feature as a major concern of the advocates for new forms of science; the latter seem more interested in notions of the best science for "representing reality" in a general, abstract way. Yet a more pragmatic or relativist stance would question whether science can or should be directed towards a single ambition or ideal.

Discussion: multiple stories, reduced realities

A major argument here is that there is no single way of describing the relation between environmental science and policy making. The four different approaches discussed differ in the extent to which they accept that point; scientism deals in an unproblematic but highly reduced reality, whereas sociology of science tends to problematize and unpick multiple constructions of reality and to aim for holistic understandings. The other two approaches are agnostic, and sometimes schizophrenic, on these issues. Many policy actors use more than a single type of argument in different contexts. For example, one research manager expressed the following views: GCMs should be the key input to climate impacts research (scientism); the satellite lobby had used scientific arguments opportunistically to appropriate resources (science as politics); and the arguments concerning climate change of the "contrarians", as well as many environmentalists, should be treated with much incredulitity (scepticism).

What, however, are the implications of these arguments for climate change research and policy? Perhaps the most important consequence will be the increasing tensions in an account of global environmental problems founded on scientific certainties. The universal claims of science are especially attractive in the context of the globalization of economic, political and cultural systems. The pressures towards standardization of science, and its relation to policy, will be undermined, however, by the multiple ways in which the science policy relationship is interpreted and used by policy actors as well as academics.

On a more prescriptive level, it seems to me that claims about "good", or "superior" science, or "privileged knowledge", should be subject to critical scrutiny and, where appropriate, challenged. In practice, this means examining the arguments and values constituting scientism, given the latter's unstated dominance in many science and policy circles. To be consistent the claims made by those developing new forms of science to produce a richer understanding should also be studied, though sympathetically. This approach should hopefully have the effect of strengthening new forms of understanding, and encourage exploration of the way in which knowledge is, and can be, used in policy making. In time, we might even want to use that experience to fashion our knowledge "through" supporting those institutions, local cultures, values and commitments which favour particular knowledge styles.

23

Notes

1. This is by no means an exhaustive list. Other approaches include discourse analysis, ethnomethodology and rational choice theory.

2. The notion that science *should* be used introduces an inescapable moral dimension into scientism, see Midgley, M. (1992), although some other authors (e.g. Barnes 1985) call this aspect technocracy.

3. Minutes, House of Lords (1994), p. 15.

4. See Boehmer-Christiansen (1994).

5. Dewey (1994), p. 7.

6. Cmnd. 137 (1993), p. 5. The following account is based on my observations at the IPCC meeting.

7. I refer here only to the drafts of the agenda as discussed at the meeting in December 1993.

8. See, for example, Midgley (1992), Sorell (1991), Barnes (1985) and Cameron & Edge (1979).

9. Le Roy Ladurie (1971).

10. ibid. p. 309.

11. e.g. Stehr & von Storch (1994), Glanz et.al. (1988).

12. e.g. Boehmer-Christiansen & Skea (1991) for the case of gas flue desulphurisation in coal-powered electricity plants. See also Weale (1992).

13. e.g. Schneider (1990), Ungar (1992).

14. Parikh (1992).

15. Even these cannot be presumed to be infallible to critical scrutiny; a wide range of assumptions and judgements goes into constructing the global temperature series of the last 100 years, for example, the validity of which can never be 100 per cent assured. It would be better to state there there are beacons of certainty within a particular framing of knowledge creation and evaluation.

16. Oreskes et.al. (1994) note that "validation" of models of open environmental systems is an impossibility; they suggest use of the term confirmation.

17. Two models which simulate the current climate equally well can produce very different responses to a prescribed doubling of CO_2.

18. See, for example, Stone (1992), Bengtsson (1992).

19. Benedick (1991), Haas (1990), Boehmer-Christiansen (1994). Negotiations on the Climate Convention are very much imbued with political, economic and ethical issues, e.g. on the issue of "joint implementation", Akumu (1992). The notion of "epistemic communities", and the role of scientific knowledge in solidifying such institutions, is relevant to this issue.

20. cf. Midgley (1992).

21. In questions session of a public lecture by Sir Crispin Tickell, Lancaster University, 17th June 1994.

22. Interview 19th July 1994, London.

23. The notions of "immutable mobiles" and "black-boxes" are from Latour (1986).

24. For criticisms and discussions of the "interests model" see Schwarz & Thompson (1990), Cambrosio et.al. (1990, 1991), and on-going debate in Kleinman (1991), Wynne (1992) and Abraham (1994). On ambivalence see Wynne (1992a). The point is not that there are no interests, but that they cannot be presumed or over reified.

25. This is not to suggest, however, that there is a single, or correct way to represent scientific knowledge and practice. The contrarians acceptance of financial support from the fossil fuel industries is a further reason to be somewhat suspicious as to their motives.

26. The presentation was by Patrick Michaels at a meeting in Phoenix, Arizona, on "Global Climate Change", April 5-8th 1994. The issue is not as cut and dried as suggested here, however, since some modellers *do* attempt some comparison of transient GCM output with observational time-series.

27. This view is illustrated particularly clearly in the writings of the late Aaron Wildavsky, a political scientist who cultivated relationships with the contrarians, including writing a foreword for Robert Balling's book *The Heated Debate* (1992). Wildavsky's use of Mary Douglas's cultural theory is a useful heuristic; see also Steve Rayner (n.d.), and Nierenberg (1993).

28. See, for example, Collins (1992 [1985]), Latour & Woolgar (1979).

29. Some of these issues are discussed in Pickering (1992).

30. This is based on the analysis in a Working Paper (Shackley et.al. 1993).

31. Mulgan (1994), p.16.

32. Compare this to Crispin Tickell's comment that: "Such action [pro-active policy] might indeed be created by the politics of fear. Whether this is the right or desirable way to bring about change I leave to you; it is often the only way that change of a radical kind comes about" (Tickell 1994).

33. Tennekes (1994).

34. Tennekes (1994a).

References

Abraham, J. (1994), "Interests, Presuppositions and the Science Policy Construction Debate", *Social Studies of Science*, vol.21, pp.123-32.

Akumu, G. (1994), "Mitigation Strategy or Hidden Agenda?", *TIEMPO*, No.11, May.

Balling, R. (1992), *The Heated Debate*, Pacific Research Institute for Public Policy, San Franciso, California.

Barnes, B. (1985), *About Science*, Blackwell, Oxford.

Benedick, R. (1991), *Ozone Diplomacy*, Harvard University Press, London.

Bengtsson, L. (1992), "Climate system modeling prospects", in Trenberth, K. (ed.), *Climate System Modeling*, Cambridge University Press, Cambridge.

Boehmer-Christiansen, S. (1993), "Science Policy, the IPCC and the Climate Convention: The Codification of a Global Research Agenda", *Energy & Environment*, vol.4, no.4, pp. 362-407.

Boehmer-Christiansen, S. & Skea, J. (1991), *Acid Politics*, Belhaven Press, London.

Cambrosio, A., Limoges, C. & Pronovost, D. (1990), "Representing Biotechnology", *Social Studies of Science*, vol.20, pp. 195-227.

Cambrosio, A. et.al (1991), "Reply to Kleinman", *Social Studies of Science*, vol. 21, pp. 775-781.

Cameron, I. & Edge, D. (1979), *Scientific Images and their Social Uses* , Butterworths, London.

Collins, H. (1992 [1985]), *Changing Order*, Chicago University Press, London.

Cm 2137 (1993), United Nations Framework Convention on Climate Change, Miscellaneous Series No. 6, HMSO, London.

Dewey, D. (1994), "Cool headed approach to a warm subject", Profile in *Baptist Times*, June 2nd, p.7.

Glanz, M. ed. (1988), *Societal Responses to Regional Climatic Change: Forecasting by Analogy*, Westview, Boulder, CO.

Haas, P. (1990), *Saving the Mediterranean*, Columbia University Press, NY.

IPCC (1990), *Climate Change: The IPCC Scientific Assessment*, Cambridge University Press, Cambridge

Kleinman, D. (1991), "Conceptualising the Politics of Science", *Social Studies of Science*, vol.21, pp.769-74.

Latour, B. & Woolgar, S. (1979), *Laboratory Life*, Sage, London.

Le Roy Ladurie, E. (1971), *Times of Feast, Times of Famine*, Doubleday, New York.

Midgley, M. (1992), *Science as Salvation*, Routledge, London.

Minutes of Evidence Taken Before the Select Committee on Sustainable Development, Tuesday 3rd May, 1994, HL Paper 66-ii, HMSO, London.

Mulgan, G. (1994), "Green wave that doesn't amount to a sea change", *The Guardian*, 2nd August, p.16.

Nierenberg, W.A. (1993), "Science, Policy and International Affairs: How Wrong the Great Can Be", *Environmental Conservation*, vol. 2, No. 3, pp.195-197.

Oreskes, N., Shrader-Frechette, K. & Belitz, K. (1994), "Verification, Validation, and Confirmation of Numerical Models in the Earth Sciences", *Science*, Vol.263, pp.641-646.

Parikh, J.K. (1992), "IPCC strategies unfair to the South", *Nature*, vol.360, pp.507-508.

Pickering, A. (1992), *Science as Practice and Culture*, Chicago University Press, London.

Schneider, S. (1990), *Global Warming*, Lutterworth Press, Cambridge.

Schwarz, M. & Thompson, M. (1990), *Divided We Stand*, Wheatsheaf, Hemel Hempstead.

Shackley, S., Wynne, B., Parkinson, S. & Young, P., "Mission to Model Earth", CSEC/CRES Working Paper, 1993.

Sorell, T. (1991), *Scientism*, Routledge, London.

Stehr, N. & von Storch, H. (1994), "Climate Change, the Social Construct of Climate and Policy", Proceedings of the 5th Symposium on Global Change Studies, Nashville, USA, Jan. 23-28th, 1994.

Stone, P. (1992), "Forecast Cloudy: The Limits of Global Warming Models", *Technology Review*, vol.95, pp.32-40.

Tennekes, H. (1994), "The Limits of Science", MS.

Tennekes, H. (1994a), personal communication, June 1994.

Tickell, C. (1994), personal communication, 23rd June.

Ungar, S. (1992), "The rise and (relative) decline of global warming as a social problem", *The Sociological Quarterly*, Vol.33(4), pp.483-501.

Weale, A. (1992), *The new politics of pollution*, Manchester University Press, Manchester.

Wynne, B. (1992), "Representing Policy Constructions and Interests in SSK", *Social Studies of Science*, vol.22, pp.575-580.

Wynne, B. (1992a), "Misunderstood misunderstanding: social identities and public uptake of science", *Public Understanding of Science*, vol.1, pp.281-304.

Wynne, B. (1994), "Scientific Knowledge and the Global Environment", in Redclift, M. & Benton, T. (eds), *Social Theory and the Global Environment*, Routledge, London.

2 Agricultural biotechnology as clean surgical strike

Les Levidow

The agricultural biotechnology industry has portrayed its technical advances as essential for achieving global security. This self portrayal can serve the public relations purposes of marginalizing political opposition and minimizing state regulation. However, the industry's language is more than merely rhetorical, it diagnoses an environmental insecurity which defines the problem for biotechnologists to solve. What does its Research and Development (R&D) agenda take for granted? How does it prescribe the future?

Such questions will be answered by analysing diverse literature: industrial, scientific and critical. In so doing, this article emphasizes issues which are marginalized by the narrow remit of "safety" regulation.

Environmental insecurities

In the language of the agricultural biotechnology industry, its products will be necessary to protect the common good from environmental insecurities of three kinds: demographic, commercial and pestilential. Let us examine how these insecurities seem to demand genetic remedies.

Demographic insecurity

We must correct genetic deficiencies in order to secure and expand the food supply, according to the USA's Industrial Biotechnology Association. That is, our society has temporarily proven Malthus wrong, because "the American farmer has adopted science and technology as rapidly as it has become available, allowing farm production to outpace population growth". Consequently, "Our existence is now dependent upon fewer than 20 species of plants; we must use all available resources to assure that [those] species are genetically fit to survive under the wide range of

31

environmental extremes" (Calder, 1991, p.71). In other words, industry has promoted a genetic uniformity which makes agriculture more vulnerable to the vicissitudes of nature; this vulnerability must now be overcome by fixing the genes.

Such arguments extend familiar neoMalthusian ones: that is, agricultural yield must keep pace with the Third World's growing population in order to avert more famines. According to Britain's single largest seeds merchant, ICI Seeds, "biotechnology will be the most reliable and environmentally acceptable way to secure the world's food supplies"; it can provide essential tools for "feeding the world" (for references see Levidow and Tait, 1991). In this vein, an ad from Monsanto depicts maize growing in the desert: "Will it take a miracle to solve the world's hunger problem?"

Moreover, it is suggested that food shortages in the Third World threaten the security of the West. As one publicist argues, "We will need dramatic progress in the productivity of agriculture to limit starvation and the social chaos which overpopulation will bring..." (Taverne, 1990, p.5). Thus, by helping the Third World to increase agricultural yields, the West can protect itself from immigrant hordes and other environmental threats.

Commercial insecurities

When invoking a demographic threat, industry publicists conveniently ignore the appropriation of Third World resources for producing cashcrop exports. In their account, the problem instead appears as overpopulation and inefficient agriculture. A similar diagnosis underlies structural adjustment programmes (World Bank, 1992), which have further dispossessed and impoverished Third World populations in the name of modernizing their countries.

The pretence of increasing food production, much less "feeding the world", is belied by the R&D priorities of the biotechnology industry. Its own house journal, *Agro-Industry Hi-Tech*, has a revealing subtitle: International Journal for Food, Chemicals, Pharmaceuticals, Cosmetics as Linked to Agriculture Through Advanced Technology; this acknowledges the priority of making biological materials more plastic and interchangeable. The journal emphasizes the political context of reduced farm subsidies, which will make productivity less important in the future: "Agriculture is bound to go for more [high] value-added products, better adapted to demand from downstream industry and the consumer. Hence it is going irresistibly towards a global system where contents matter more than quantities" (Anon, 1990).

In our society, what defines "value added"? According to US Tobacco's Vice-President, "value-added genetics determines the processability, nutrition, convenience and quality of our raw materials and food products" (Lawrence, 1988, p.32). Concretely, some biotechnology companies are developing substitutes for crops or materials hitherto imported from Third World countries (Hobbelink, 1991, p.93; Walgate, 1990, p.57; Panos, 1993, pp.12-14); for example, oilseed rape is being altered to produce oils which could substitute for those hitherto imported from tropical countries. If technically successful, these new products would undermine the livelihoods of entire Third World communities, just as European sugar beet has devastated sugar cane production elsewhere.

The biotechnology industry euphemistically portrays its disruptive power as democratic progress: "Let there be no illusions: as with any innovative technology, biotechnology will change economic and competitive conditions in the market. Indeed, economic renewal through innovation is the motor force of democratic societies" (SAGB, 1990, p.15).

These competitive conditions demand more flexible investment strategies in the face of an insecure commercial environment. "Value-added genetics" directs R&D towards accommodating and aggravating commercial insecurities.

Pestilential insecurity

As chemical pesticides continue to encounter public resistance, some biotechnology R&D seeks more acceptable methods of crop protection. One line of research attempts to replace agrochemicals with new biopesticides. A gene for a toxin can be transferred from a microbe to another organism which will persist longer or target the pest more effectively. Some R&D even combines genes for different toxins in the same microbe, thus killing a broader range of insect pests.

What problem is this solving? Traditional biopesticides have a narrow host range, which limit their commercial potential. Laments specialist Sheldon Murphy, "It is like a rifle shot rather than a shotgun shot into a pest group" (quoted in Kloppenburg, 1988, p.251). Microbial pesticides occupy only about one per cent of the pesticide market, the small share due to their "lack of environmental persistence, narrow host range, limited virulence, and high production costs" (Cook and Granados, 1991, p.217; cf. McManus, 1989, p.65).

Those features, which make biopesticides so attractive ecologically, also make them unattractive economically to companies, regardless of whether the products would be attractive to farmers. As the solution, a genetic redesign can make the biopesticide less specific, more persistent and/or more deadly. In those ways, biotechnology may overcome the economic limitations of traditional biopesticides. Thus "value-added genetics" predefines the pestilential insecurity to be overcome.

Cleaner defence?

In strengthening genetic defences, biotechnologists reconceptualize the type of "clean" nature which will make agriculture safe from pests. Let us examine the conceptual shift.

In the industrialized agriculture which prevailed after World War II, plant breeders could select crop strains mainly for high yield, shielded by the "pesticide umbrella"; chemicals protected crops from insects, weeds and disease. Farmland was kept "clean" of intruders and indeed, of all other life. The soil became less able to regenerate itself; its fertility came to depend upon applying chemical fertilizer rather than recyling vegetation or manure. As a cultural critic has observed, "That ultimate simulacrum of our times -- artificial shit -- is surely the sign of a culture obsessed with what Baudrillard calls 'deadly cleanliness'" (Nelson, 1990).

Since the 1960s, the chemicals have been losing both their clean image and agronomic effectiveness. Genetic uniformity leaves crops more vulnerable to pests and disease. Pesticides eliminate the natural predators of pests and/or generates selection pressure for insect pests resistant to the chemicals. Agriculture faces a "chemical treadmill", needing new pesticides to keep up with new pest resistance.

Despite applying more and newer pesticides, agriculture has suffered even greater crop losses; widely attributed to intensive monoculture methods, which abandoned such practices as crop rotation (e.g. Pimentel, 1989, p.70). More and more fertilizer is been needed to sustain crop yields, yet fertilizers assist weeds. Political protest has cited threats to human and environmental safety from chemical residues or run off, in turn leading government to restrict pesticide use.

Recently agronomists have been acknowledging the limits of overcoming pest problems through better chemicals alone. However, industry now tends to locate the problem within genetic deficiencies, which can be corrected by inserting extra genetic defences into crops or biopesticides. In this way, biotechnological methods are accorded a natural legitimacy. "What is new is our growing ability to simulate nature in ways that can offer enormous benefits", declares the President of Mycogen Corporation (Calder, 1991, p.75; cf. Goodman, 1989, p.49).

Within agricultural biotechnology, the greatest R&D efforts are directed at developing herbicide resistant crops. Their inserted gene protects the crop from broad spectrum herbicides, which in turn kill all other vegetation. Previously agronomists had to find herbicides which would selectively

34

spare the crop from damage; now the inserted gene provides the selective protection.

By inserting a gene which offers resistance to a less persistent herbicide, industry can describe the product as cleaner, in the dual sense of combining a precise defence with a less polluting chemical. According to ICI Seeds, herbicide resistant crops will reduce dependence upon chemical inputs, by reducing quantities of their use; such crops will offer greater choice to farmers, who can thereby defer herbicide applications until the post emergence phase (e.g. Bartle, 1991). These products have even been celebrated as the ultimate solution to the problem of weeds resistant to herbicides, even as a "moral imperative" for feeding the Third World (Gressel, 1992, 1993).

Critics regard this R&D priority as perpetuating dependence upon chemical herbicides, regardless of whether quantities are reduced (BWG, 1990). Ecologists remind us that some herbicides have already weakened crops' natural defences and so necessitated additional chemical treatments (Pimentel, 1987). In various ways, herbicide resistant crops could extend the use of agrochemicals, across time and space.

In this vein, imagine a "supercrop" which has been genetically modified for resistance to both insect attack and high doses of herbicide. Such an imaginary crop is depicted in a textbook for schoolchildren, sponsored by Britain's Department of Trade and Industry (Satelle, 1988, p.31). This image exemplifies the declaration by an industry publicist: "if we have the imagination and resources, there is almost no biological problem we cannot solve" (Taverne, 1990, p.4).

Not merely rhetorical, such fantasy has roots in the conceptual framework of biotechnology: reprogramming nature for total environmental control. In the 1930s, the new science of molecular biology treated genetic material as interchangeable, universal coded "information", which would permit the ultimate human control over the "essence of life" (Yoxen, 1983). With the cell now conceptualized as a natural factory, the "factory farm" also becomes more than a metaphor (Krimsky, 1991, p.10).

Indeed, genetic engineering invests nature with computer and industrial metaphors, which in turn lend a natural status to its products. Emphasizing the universal genetic code in nature, Monsanto presents genetic engineering as a "natural science". Paradoxically, biotechnology does what nature does; but does so more safely and efficiently (Kleinman and Kloppenburg, 1991).

In the late 1980s, the prevalent language shifted from genetically engineered to modified organisms (GMOs), the new term denoting a modest improvement upon nature. Genetic modification precisely enhances natural characteristics; for example, by "giving nature a little nudge towards greater efficiency", according to ICI. Akin to nature, and protective of nature, GMOs can appear to provide an enhanced natural efficiency through "environment friendly products" (Levidow and Tait, 1991). Through this rhetorical greening, biotechnology can be promoted as "clean technology".

Its R&D seeks a total biosystems control by manipulating a few genetic and/or chemical parameters (Kloppenburg, 1991). Like the strategy of purely chemical control, this genetic level control treats the systemic instabilities of intensive monoculture as external natural threats. Deploying a precise genetic defence, possibly combined with a broad spectrum chemical offence, biotechnology offers a clean surgical strike against unruly nature.

DNA as chemical bullets

Within this biotechnological perspective, how are genetic resources appropriated for crop protection? Let us examine expert disagreements over R&D strategy for new biopesticides. (See also the earlier section on pestilential insecurity.)

A traditional biopesticide, Bacillus thuringiensis, has been widely used in agriculture. Commercialization expanded in the late 1960s, after isolating a Bt strain particularly effective against Lepidopteran pests (McManus, 1989, pp.62-63). Bt is also the potential basis for new biopesticides, as it has numerous varieties, each of which produces a toxin specific to certain insects. The corresponding genes are being identified and transferred into more persistent organisms, be they other micro organisms or even plants.

Although sceptical of simple solutions, one entomologist speaks in military metaphor: each Bt toxin is "like a surgical tool for taking out the pest" (Fred Gould in Holmes, 1993, p.34). Emphasizing the selective precision, one biotechnologist notes that genetic modification provides the plant breeder with "more ammunition to help him hit his target" (Lindsey, 1991, p.9). According to a former Vice-President of Calgene, biotechnology attempts "to do better than mother nature in designing improved, more efficacious toxins" (Goodman, 1989, p.52). Novo Nordisk, another leading company in biopesticides, portrays their benign surgical precision with the visual metaphor of a green bow and arrow: "Fighting for a better world, naturally".

Regardless of their precision and genetic origin, the new biopesticides pose a familiar hazard: generating selection pressure for resistant pests. If such resistance weakens the effectiveness of traditional biopesticides as well as the new one, then it will eliminate a relatively safe alternative to chemicals, as some environmentalists have warned. Such warnings have been strengthened by scientific reports that some pests are developing resistance to Bt in stored grain and even to Bt in the field. As a journalist noted, "Mother Nature has startled the genetic scientists..." (Connor, 1991), though they were startled only because they had a tunnel vision.

To avoid pest resistance, scientists have been discussing strategies for integrated pest management (IPM). In the case of chemical pesticides, even a leading advocate of IPM has suggested that it "will only slow the pesticide treadmill, thereby extending the usefulness of available chemicals" (Hammock and Soderlund, 1986, p.113). By analogy, can genetically modified biopesticides be designed to avoid a "genetic treadmill", or at least to slow it down?

Many scientists warn that a genetic treadmill will result from any attempt at totally exterminating a pest. They seek alternative strategies which can "outwit evolution", that is by minimizing selection pressure for resistant pests. Early in the Bt debate, a Calgene official suggested using genetically modified Bt "to control [insect] populations rather than kill insects outright" (Goodman, 1989, p.52).

Some strategies, amenable to IPM, have been proposed by entomologists (e.g. Gould, 1988) and endorsed by some industrialists (Goodman, 1989; Cutler, 1991). On the military analogy of a "safe haven", farmers could provide refugia, that is patches of toxin free plants, so that the less resistant insects can survive and pass on their genes; and/or farmers could vary the choice of toxin in space and time. Yet even proponents of the refugia strategy acknowledge great difficulties in designing and implementing it, partly because the necessary measures would impose commercial disadvantages upon farmers and/or seed vendors (Holmes, 1993). The strategy assumes a willingness to sacrifice economic benefits, in favour of the longer term common good.

Attempts at IPM may benefit from new genetic knowledge. At Plant Genetic Systems, laboratory research has clarified that pest resistance to the different Bt toxins is controlled independently, by different genes (Peferoen, 1991; van Rie, 1991). Such knowledge implies that varying the toxin over time could help prevent pest resistance. However, one PGS researcher downplays IPM strategies, instead arguing that "insecticide mixtures would be more effective": this strategy assumes that no insects would be resistant to more than one toxin. He concludes that "the optimal strategy for pest management will depend upon the genetic basis for resistance" (van Rie, 1991, p.179).

His more cautious colleagues at PGS foresee new biopesticides bringing a massive introduction of Bt in soils; they warn that "we know almost nothing about its ecology" (Lambert and Peferoen, 1991, pp.120-21). Moreover, the strategy of mixing toxins would have to be executed perfectly in order to succeed; otherwise it would generate multiply resistant pests; a "superpest", as a Du Pont scientist warns

(Holmes, 1993, p.36). Additional field testing has shown that moderately resistant insects can evolve resistance to Bt mixtures (Tabashnik et al., 1991); thus such a product seems likely to generate a genetic treadmill.

Nevertheless, the prevalent R&D treats genes as chemical bullets for a total extermination strategy; this conceptualizes crop vulnerability as an external problem of pest genetics. For example, biotechnologists develop "cassettes" for inserting several defence genes at once into a plant or micro organism (Day, 1993, p.39). This technique happens to coincide with commercial pressures for mixing different toxins in the same product. According to USDA scientists, a refugia strategy is probably better for avoiding resistance, but simpler strategies, such as toxin mixtures, would be more profitable for industry (McGaughey and Whalon, 1992, p.1455).

What if a "genetic treadmill" ensues? According to Jerry Caulder, President of Mycogen Corporation, insects have understandably acquired resistance to the one strain of Bt which has been used for thirty years, but "We have other bullets in the gun we call Bt" (Cutler, 1991). From this cornucopian perspective, any one toxin is dispensable, because scientists will always find another one to kill the same pest. Indeed, some researchers are already planning how to add new toxins faster than the insect pests can develop resistance to the previous ones (Wilson et al., 1992). Thus a prospective "genetic treadmill" is treated not as a hazard, but rather as a challenge for technical advance and commercial advantage. (By analogy to the New World Order, a global control system manages and intensifies its own endemic instabilities, while attributing these to an external enemy, from which we need perpetual protection; see Levidow, 1994).

Total control?

In conclusion, biotechnology presents itself as a naturally based alternative to agrochemicals, yet its dominant paradigm conceptualizes DNA as the ultimate chemical weapon. As an historical precedent, medicine too has appropriated military metaphors in developing new weapons which can seek and destroy invading pathogens (Montgomery, 1991). Both the medical and military metaphors converge in biotechnology: here DNA becomes a magic bullet for "cleaning up on the farm", in both senses of that verb (Levidow, 1991a).

The term "biodiversity" comes to mean new combinations of special protective genes in crop strains. Traditionally, diverse cultivars (and biopesticides) provided a systemic defence against unanticipated pests or disease. This biodiversity now becomes a resource for genetic prospecting, for extracting a few magic bullets which can provide high value-added commodities.

An organic farmers' organization has warned that "supercrops" may encourage farmers to buy seed anew each season rather than sow seed harvested from the previous year's crop. The resulting "global monoculture" will become all the more vulnerable to pests and disease (Bill Duesing, quoted in Day, 1993). These problems derive from our intensive monocultural system, driven by imperatives of profitability, as newly embodied in biotechnology R&D agendas (e.g. Kloppenburg, 1988; Hobbelink, 1991).

Toward a sustainable agriculture, critics counterpose "holistic" methods of crop protection, such as those which have traditionally helped to avoid weeds, pests and disease: "It is widely agreed that systems approaches -- for example, crop rotation and other methods -- could avoid the need for the majority of pesticides, both chemical and biological, now and into the future" (Mellon, 1991, p.67).

Some alternatives idealize traditional methods as proximate to nature: "The closer a farming system comes to a natural ecosystem, the most likely it is to be sustainable" (Hobbelink, 1991, p.140). On this model, society can appropriate benign ecological processes and so keep the agricultural system clean of artificial contaminants, that is high-tech inputs.

In contending versions of "clean" agriculture, protagonists tend to idealize or demonize some external Nature. At the same time, such concepts mediate a struggle over farmers' social power. Industry emphasizes that modern pest control methods can reduce farm management time (e.g. Goodman, 1989, pp.84-86). Yet critics foresee biotechnological methods dispossessing farmers of their knowledge and skills, which alternative methods could strengthen (Hassebrook, 1989). In pest control, for example, a total extermination strategy would initially spare too few insects to warrant monitoring them, thus discouraging the skills associated with IPM.

Also at issue is the meaning of "sustainable agriculture", which concerns more than simply quantitative yields. For example, does "sustainable agriculture" signify only an environmental equilibrium or also a way of life? (See MacDonald, 1989, p. 20; Hamlin, 1991.) The question could be extended, by asking how any model, of equilibrium or "clean" agriculture, tends to favour one way of life over another.

Amidst these debates, alternative approaches have obtained little research funding. US government funds have been shifting towards "biotechnological product development, rather than biological process understanding" (Doyle, 1990, p.191). As a leading biotechnologist proclaimed, "This is the era of biology, and we are the biologists": meaning not ecologists or even agronomists (cited in Levidow, 1991b).

The predominant R&D strategy constructs a good, biotechnologized Mother Nature against the external threat of a bad, unruly nature. It invests nature with metaphors of codes, commodities and combat. This search for a clean surgical strike will prevail, unless opposition forces can reconstitute agriculture according to different social metaphors.

Acknowledgements

This essay arises from a research project, "Regulating the Risks of Biotechnology", funded by the Economic and Social Research Council (project number R000 23 1611). I would like to thank David Pimentel and Roger Wrubel for helpful editorial comments.

References

Anon (1990), "Editorial", *Agro-Industry High-Tech* Vol.1(1), pp.3-4, Teknoscienze, Milano.

Bartle, I. (1991), *Herbicide-Tolerant Plants: Weed Control with the Environment in Mind*, ICI Seeds, Haslemere, Surrey.

Beck, U. (1992), *Risk Society: Towards a New Modernity*, Sage, London.

BWG (1990), *Biotechnology's Bitter Harvest: Herbicide Tolerant Crops and the Threat to Sustainable Agriculture*, Biotechnology Working Group, available from Environmental Defense Fund, 257 Park Avenue South, New York, NY 10010 (price $10).

Calder, J. (1991), "Biotechnology at the Forefront of Agriculture", pp.71-75 in (ed.) MacDonald, op. cit.

Cook, R. and Granados, R. (1991), "Biological Control: Making It Work", pp.213-227 in (ed.) MacDonald.

Connor, S. (1991), "Mother Nature Thwarts Hopes of Supercrops", *Independent on Sunday, 17 November*.

Cutler, K. (1991), "Bt Resistance: a Cause for Concern?", *Ag Biotech News*, January/February, p.7.

Dart, E. (1988), *Development of Biotechnology in a Large Company*, ICI External Relations Department [now Zeneca], London.

Day, S. (1993), "A Shot in the Arm for Plants", *New Scientist*, 9 January, pp.36-40.

Doyle, J. (1990), "Who Will Gain From Biotechnology?", pp.177-193, in (eds) S. Gendel et al., *Agricultural Bioethics: Implications of Agricultural Biotechnology*, Iowa State University, Ames, IA.

Goodman, R. M. (1989), "Biotechnology and Sustainable Agriculture: Policy Alternatives", pp.48-57 in MacDonald.

Gressel, J. (1992), "Genetically-Engineered Herbicide-Resistant Crops: A Moral Imperative For World Food Production", *Agro-Food-Industry Hi-Tech*, Vol.3(6), pp.3-7, Teknoscienze, Milano.

Gressel, J. (1993), "Herbicide Resistance Threatens Wheat Supply", *Agro-Industry Hi-Tech*, Vol.4(2), p.36, Teknoscienze, Milano.

Gould, F. (1988), "Genetic Engineering, Integrated Pest Management and the Evolution of Pests", pp.15-18 in (eds) Hodgson, J. and Sugden, A.M., *Planned Release of Genetically Engineered Organisms*, (TREE/Tibtech), Elsevier, Cambridge.

Hamlin, C. (1991), "Green Meanings: What Is Sustained By Sustainable Agriculture?", *Science as Culture*, Vol.2(4), pp.507-537, Free Association Books, London/Guilford Publs, New York.

Hammock, B. and Soderlund, D. (1986), "Chemical Strategies for Resistance Management", pp.11-29 in National Research Council, *Pesticide Resistance: Strategies and Tactics for Management*, National Academy Press, Washington, DC.

Hassebrook, C. (1989), "Biotechnology, Sustainable Agriculture and the Family Farm", pp.38-47 in MacDonald.

Hobbelink, H. (1991), *Biotechnology and the Future of World Agriculture*, Zed, London.

Holmes, B. (1993), "The Perils of Planting Pesticides", *New Scientist* 28 August, pp. 34-37.

41

Kleinman, D.L. and Kloppenburg, J. (1991), "Aiming for the Discursive High Ground: Monsanto and the Biotechnology Controversy", *Sociological Forum*, 6 (3).

Kloppenburg, J. (1988), *First the Seed: The Political Economy of Plant Biotechnology, 1492-2000*, Cambridge University Press, Cambridge.

Kloppenburg, J. (1991), "Alternative agriculture and the new biotechnologies", *Science as Culture*, Vol.2(4), pp.482-506, Free Association Books, London/Guilford Publs, New York.

Krimsky, S. (1991), *Biotechnics and Society: The Rise of Industrial Genetics*, Praeger, London.

Lambert, B. and Peferoen, M. (1992), "Insecticidal promise of Bacillus thuringiensis", *BioScience*, Vol.42(2), pp.113-121.

Lawrence, Robert H. (1988), "New Applications of Biotechnology in the Food Industry", pp.19-45 in *Biotechnology and the Food Supply*, National Academy Press, Washington, D.C.

Levidow, L. (1991a), "Cleaning Up on the Farm", *Science as Culture* Vol.2(4), pp.538-568, Free Association Books, London/Guilford Publs, New York.

Levidow, L. (1991b), "Biotechnology at the Amber Crossing", *Project Appraisal*, Vol.6(4), pp.234-238.

Levidow, L. (1994), "The Gulf Massacre as Paranoid Rationality", in G. Bender and T. Druckrey (eds), *Culture on the Brink: Ideologies of Technology*, Bay Press, Seattle.

Levidow, L. and Tait, J. (1991), "The Greening of Biotechnology: GMOs as Environment-Friendly Products", *Science and Public Policy*, Vol.18(5), pp.271-280.

Lindsey, K. (1991), "Crop Improvement Through Biotechnology", *Agro-Industry Hi-Tech*, Vol.2(4), pp.9-16.

McGaughey, W., and Whalon, M. (1992), "Managing Insect Resistance to Bacillus thuringiensis Toxins", *Science*, Vol.258, pp.1451-55.

MacDonald, J.,(ed.) (1989), *Biotechnology and Sustainable Agriculture: Policy Alternatives*, NABC, 211 Boyce Thompson Institute, Tower Road, Ithaca, NY 14853-1801.

MacDonald, J. (ed.) (1991), *Agricultural Biotechnology at the Crossroads: Biological, Social and Institutional Concerns*, NABC, 211 Boyce Thompson Institute, Tower Road, Ithaca, NY 14853-1801.

McManus, M. (1989), "Biopesticides: An Overview", pp.60-68 in MacDonald.

Mellon, M. (1991), "Biotechnology and the Environmental Vision", pp.66-70 in MacDonald.

Montgomery, S. (1991), "Codes and Combat in Biomedical Discourse", *Science as Culture* 2 (3), pp.341-90.

Nelson, Joyce (1989), "Culture and Agriculture: The Ultimate Simulacrum", *Border/Lines* spring, pp. 34-38, Bethune College, North York, Ontario.

Panos (1993), *"Genetic Engineers Target Third World Crops"*, Media Briefing no.7, December, Panos Institute, London.

Peferoen, M. (1992), "Bacillus thuringiensis in Crop Protection", *Agro-Industry Hi-Tech* 2(6), pp.5-10.

Pimentel, D. (1987), "Down on the Farm: Genetic Engineering Meets Ecology", *Technology Review*, January, pp. 24-30.

Pimentel, D. (1989), "Biopesticides and the Environment", pp.69-74 in MacDonald.

SAGB (1990), *Community Policy for Biotechnology: Priorities and Actions*, Senior Advisory Group on Biotechnology/CEFIC, Brussels.

Satelle, D. (1988), *Biotechnology... in Perspective*, Hobsons, Cambridge.

Tabashnik, B., Finson, N. & Johnson, M. (1991), "Managing Resistance to Bacillus thuringiensis: Lessons from the Diamondback Moth (Lepidoptera: Plutellidae)", *Jnl Econ. Entomology*, Vol.84(1), pp.49-55.

Taverne, D. (1990), *The Case for Biotechnology*, Prima, London.

43

van Rie, J. (1991), "Insect Control with Transgenic Plants: Resistance Proof?", *Tibtech*, Vol.9, pp.177-179

Walgate, R. (1990), *Miracle or Menace? Biotechnology and the Third World*, Panos, London.

Wilson, F.D. et al. (1992), "Resistance of Cotton Lines Containing a Bt Toxin to Pink Bollworm and Other Insects", *Jnl Econ Entomology*, Vol.85, pp.1516-1521.

World Bank (1992), *Development and the Environment*, Washington, D.C.

Yoxen, E. (1983), *The Gene Business: Who Should Control Biotechnology?* Pan, London, reissued 1986, Free Association Books, London.

3 Threats and defences in the built environment

Elizabeth Shove

In popular imagination, the environment is "everything out there": it is the outdoor world of fields and trees, birds and bees. Actions and policies for environmental protection implicitly relate to this domain, some focusing upon the intrinsic qualities of nature, others viewing the environment as a finite stock of natural resources, or as an endangered and fragile system. This, it seems, is the world which is to be saved and valued. It is important, however, to acknowledge that this is not the place in which we routinely live. Athough whales and the ozone layer are critical elements in the environmental iconography, this outside world is not the one which most people inhabit most of the time. The environment of everyday life, is, of course, the indoor environment. Current definitions of the environment as "everything out there" place people in a specific relationship to nature, setting them apart from the real environmental action. This has not always been the case for meanings of "out there" and "in here" have changed over time. This chapter examines a few moments in the history of this evolving relationship and in doing so provides a new way of looking at familiar debates about nature and culture, about science, technology and the mastery of nature, and, most recently, about theories of environmental change and strategies for minimizing environmental damage.

Keeping the elements at bay

Let us start at the beginning of human history. People are rather fragile creatures: they die if they get too hot or too cold or too wet or too dry, and they are generally unable to survive over exposure to the natural environment. Body temperature must be maintained within certain limits and anyone living outside Mesopotamia, where the outdoor environment needs no modification, has to take action to protect themselves from the outside world.

Protective clothing represents one such defensive strategy, but it is often more effective to create rather larger pockets of habitable space by constructing some form of shelter: by building an igloo, making a mud hut, a log cabin or a thatched cottage. These more accommodating structures provide a first line of defence against the hostile external environment. Other strategies, such as the introduction of some form of heating, cooling or ventilation further modify the outdoor climate. Thus liberated from the immediate demands of outdoor survival people are free to move around at will, meeting and engaging with each other whatever the weather outside.

Seen as systems for keeping the environment at bay, buildings have different jobs to do depending upon the precise nature of the outdoor climate in different parts of the globe. In some places very thick walls are required, whilst in others the thin skin of a tent does the job. Details of traditional building construction vary according to regional climate conditions as well as other factors such as the local availability of alternative building materials. Timber, for instance, has been popular in Scandinavia, stone in the Outer Hebrides, snow and ice in the Arctic. By these means the reach of habitable territory has extended around the globe. While basic survival is possible just about anywhere, efforts to create comfortable indoor environments, even when supplemented by systems of heating and cooling, have not always been entirely effective. In Northern Europe the changing seasons had a direct impact on daily life. As the following description of French peasant life at the beginning of the nineteenth century illustrates, people really felt the cold of winter: "In the Planèse, where there is absolutely no wood, the peasant would be horribly miserable during the winter and could not live there if he had not discovered the means to do without wood to get warm: he does it by living in the midst of his farm animals" (Flandrin, 1979, p.107). In this respect, the daily life of nineteenth century French peasants was closely "in tune" with nature and the seasons of the year.

The development of more effective heating and ventilating technologies has taken place within, and has in turn affected the social organization of any number of different societies. In Sweden, those who could afford a *kakelugn* (an elaborate stove in which smoke from an enclosed fire is forced to circulate within a massive ceramic covered chimney) were able to live year round in parts of the country which were previously uninabitable. There are many other such examples for methods of climate control which generally have direct and immediate consequences for individual and collective lifestlyes.

46

Differential access to more and less effective means of indoor climate control frequently mirrors wider patterns of social inequality. Not everyone can head up to the hills for the hot season and not everyone has a summer house to escape to. In colder climates, people who suffer from fuel poverty (Boardman, 1991) do not experience winters in the same way as those who are able to enjoy the comforts of full central heating. As these brief examples suggest, heating and cooling technologies reach deep into the social fabric, influencing the tiniest details of everyday life.

Taking a broader view, the capacity to exclude or minimize daily and seasonal variation in the outdoor climate in turn influences daily and seasonal patterns of social activity. In Spain, the weather penetrates the culture to such an extent that whole cities close down at 2 pm, only coming back to life again in the late afternoon. Telephones stop ringing and shutters drop in a well coordinated process of social adjustment to the natural world. While this practice is still well entrenched, retreat is no longer the only option. There are now any number of ways of artificially conditioning the indoor climate so permitting a more continuous working day. Such sociotechnical adjustment would require a substantial renegotiation of existing practices and patterns on both a societal and individual, basis. The siesta may persist for some time to come, but as others have observed, the social ramifications of air conditioning have been substantial. In the USA, cooling technologies have, for instance, "hastened the demise of the front porch, isolating families and fragmenting communities" (Morrill, 1994, p.7). As before, the social and the technical are tightly knitted together when it comes to the control and management of the indoor climate.

The science of comfort

It is important to note that the technologies of environmental exclusion have changed substantially in the last hundred years or more. Building materials have become commodities to be measured, valued, bought, sold and moved around the world, and new materials such as mineral fibre insulation, plasterboard, aluminium and manufactured brick have been developed. These materials are produced and used in such climatically varied locations as Singapore, Stockholm and Saudi Arabia. In similar fashion, heating and ventilating technologies have developed and standardizsed beyond all recognition. New industries have grown up to condition the very air we breathe.

Equally importantly, the indoor climate has come in for a massive dose of scientific analysis and investigation. People, treated here as physiological beings, have been subjected to experimental tests and measurements in a concerted effort to define the critical variables of comfort. Reaching beyond the minimum conditions for human survival, researchers have sought to identify the "optimum conditions for people engaging in activities within buildings" (Rubin and Elder, 1980, p.185). Often allied to an interest in work performance, these studies tend to concentrate on quantifiable features like body and skin temperature, relating these and other "objective" measures of temperature, humidity and air velocity to people's experiences of comfort. More recent investigations focus upon similar themes, work by the National Swedish Institute of Building Research suggesting that "productivity falls by 30 to 50 per cent when the temperature inside a building reaches 27° C, compared to 20° C" (Energy in Buildings and Industry, May 1994, p.6). Scientific analysis of the conditions and parameters of comfort and careful study of the relationship between comfort and performance has provided a seemingly solid basis for subsequent design decisions. Accordingly, the results of such enquiries, especially those of Fanger (1970), have fed directly into the design process. The notion of a "comfort zone", that is of a range of indoor climatic conditions which suits most people most of the time, is now firmly established within conventional design guidance. Buildings are consequently constructed, and the heating and ventilating systems specified, so as to create standardized, statistically optimal, indoor environments.

The science of comfort, linked to the internationalization of building materials and to world wide sales of heating and ventilating equipment, has stealthily brought about an indoor climate revolution of some significance. In cooler countries, the spread of full central heating is matched by the growth of air conditioning in those with hotter climates. The use of air conditioning in American homes is reported to have increased by 20 per cent since 1984 to a level of approximately 60 per cent in 1992 (Sturm, Lord and Wagner, 1992, p.48). One way or another, indoor climates are converging: hot environments are being cooled while indoor temperatures are increasing in colder parts of the world. Now that we can live in a protected bubble of comfort, now that we can be so completely cocooned by the built environment, we have no need to make seasonal adjustments to our lifestyles: there is no need to move in with the cows over winter. Nor is there any reason why we should be less comfortable inside a building in one part of the world than in another.

The all pervasive business suit provides a telling indicator of the steady homogenization of the indoor climate. Now that we have the means to exclude the outdoor environment and ignore its daily and seasonal variation people can dress the same way the world over. This technological potential generates a number of sociocultural tensions as traditional, climatically specific, ways of life bend in response to the opportunities presented by modern, climatically standardized conditions. The indoor climatic politics of the Saudi Arabian government illustrates the problem. The needs of those clad in arab clothing, the traditional dress of the Saudi government, clash with the requirements of those, expatriates and others, who expect to wear a "normal" suit (Siddiqui, 1990) when both must share the same indoor environment. Creeping standardization of the indoor climate by means of the International Standard ISO 7730 and by the routine adoption of accepted methods for calculating heating and cooling loads, has further far reaching social implications. In the banking sector, especially, work can and does continue around the clock in the air conditioned halls of high finance. Spreading outwards, offices and shopping malls acquire the same indoor climatic characteristics. Domestic properties tend to follow this trend and trains, buses and even cars now have their own systems of climatic control. As the capacity to maintain a more or less continuous thermal environment increases, so exposure to raw, unmediated, outdoor weather decreases. In the developed world, people spend, on average, between 75 per cent and 90 per cent of their life indoors, the percentage being even higher for the sick and unhealthy (Indoor Air International, 1994).

Technological efforts to manage and control the indoor environment clearly relate to changing beliefs and expectations about comfort and lifestyle. We expect more than the minimum shelter offered by the cow shed or the igloo, for now we know what comfort is, now the technologists have put a figure on it, and it has become something of a universal aspiration. The translation of techologically based standards and measurements into taken for granted social norms is a critical, if poorly understood, process. As Nick Baker wryly observes, "It could be that the very existence of definable standards for mechanically-conditioned buildings ...has been the main cause for the proliferation of air-conditioning" (Baker, 1993, p.103). By whatever means, "Comfort has changed not only qualitatively but also quantitatively - it has become a mass commodity" (Rybczynski, 1987, p.220). In this respect, the spread of full central heating within Northern Europe and of air conditioning within Southern Europe tells us as much about cultural convergence as it does about the technologies and economies of heating and cooling.

To summarize, the outdoor climate has been controlled, mastered and beaten into submission by the men in white coats and by their blue boiler suited colleagues. It has had all the unpredictability and all the variation taken out of it: so much so that buildings are perhaps better seen as machines creating their own weather than as protective systems merely modifying the external elements.

Redefining the risks

The trouble is that the manufacturing of standardized comfort comes at a price. Like other machines, buildings consume resources and as such they have potentially destructive environmental consequences. The oil crisis of the 1970s represented a critical moment in the history of the indoor climate, bringing home a sudden and shocking recognition of the costs of maintaining buildings designed as high-tech climatic fortresses. Related acknowledgement of finite fossil fuel reserves shattered the simple two part view of inside and out. It was horribly clear that the indoor environment was still firmly connected to "nature" at least in the sense that natural resources were being depleted in order to maintain standardized comfort conditions inside. By defending themselves from the elements so effectively people were beginning to endanger the sustainability of the outdoor environment. Previously seen as simple, unsophisticated defensive structures, buildings took on a new environmental role as the greedy consumers of limited resources.

The focus of concern has shifted again in recent years and arguments about the need to reduce CO_2 emissions are now as important as those relating to resource depletion, especially so since International Energy Agency figures suggest that half of all CO_2 emissions relate to the energy used in buildings. Cast in these terms, buildings and associated heating and cooling systems threaten to pollute what has come to be seen as a vulnerable and fragile natural world.

Perceptions of the relationship between inside and out have evidently shifted. The outside world is no longer characterized as a hostile and threatening domain to be excluded at all costs. Instead, it is something to be cared for, respected, and ever so carefully managed. This shifting perception has also drawn attention to links and interconnections between indoor and outdoor environments. The former is, it seems, frequently maintained at the expense of the latter.

Threats and defences in the built environment

Responding to these changing interpetations building scientists have sought to develop more energy efficient methods of producing and maintaining comfortable indoor environments. In other words they have tried to develop ways of achieving the same indoor climate but at less cost to the wider world outside. Researchers and manufacturers have, for example, produced computerized building energy management systems with sensors and controlling devices monitoring every aspect of the indoor climate in the hope of reducing "waste" by fine tuning demand. In extreme cases, "intelligent buildings" totally exclude the outside world: windows do not open, doors operate automatically and the building is sealed to permit maximum control and maximum energy efficiency. Improving technical performance is one strategy but even the most efficient heating and cooling systems function badly if located in substandard building structures. Recognizing this, efforts have also been made to develop more effective materials for insulating roof, walls and floor. Such technological solutions generally do reduce energy consumption thus helping to protect the outdoor environment. The difficulty is that these environmentally friendly materials and methods may themselves endanger the quality and safety of the indoor climate.

Take the case of cavity wall insulation. Urea formaldehyde foam, a cheap and efficient insulating material, was installed in over a million British homes before rumours of potential health scares spread. Accepting new chemical technologies without question, innocent householders suddenly discovered that they had inadvertenly injected potentially dangerous substances into the heart of their home. In the short term, formaldehyde fumes seemed to cause respiratory problems. The Pettengell family, for instance, could not "enter their bathroom without eyes streaming" (*Building Design*, June 1982). In the same year the St Thomas More school in Essex closed "as UF foam fumes fill classrooms" (*Building Design*, April 1982). In the longer term, there was some suspicion that urea formaldehyde could cause cancer. As the moral panic intensified, and as further cases of urea formaldehyde related problems came to light, government scientists were called in to establish the indoor environmental risks of this particular form of energy conservation (Shove, 1991).

According to Britain's Energy Efficiency office, cavity wall insulation represents the single most effective environmental action a householder can take. In one step it promises to reduce household energy consumption by roughly 30 per cent, with clear environmental benefits all round. Yet anxieties about the long term safety of mineral fibre and of other insulants linger on. Such generalized distrust surfaces from time to time, periodic scares about the "killer in the loft" persisting despite volumes of reassuringly scientific evidence. Materials are not the only focus of concern for there are other worries about the environmental conditions experienced in highly insulated buildings. As guidance from Britain's Building Research Establishment (BRE, 1989) emphasizes, "tightening the building envelope", keeping the outside out, must be done with care, as the unwary pursuit of energy efficiency is a potentially hazardous enterprise.

Research into health and indoor air quality is a fast developing field as the number of "sick" buildings continues to rise. The notion of sick buildings is in itself a telling indicator of the realigngment of environmental concerns, suggesting that fears about man made hazards and the manufactured risks of indoor life have upstaged historic anxieties about the trials of winter or the unsufferable heat of the mid day sun (Beck, 1992). Buildings appear to be capable of causing sickness, asthma, irritation and who knows what other harms to their unsuspecting inhabitants (Tyler, 1991). Suspicions that "the increasing prevalences of allergies and hyperactivity may to some extent be related to tight modern houses with insufficient ventilation and bad air quality" (Anderson et. al., 1992, p.1) have prompted research on a number of different fronts. Redefined as a source of potential danger, the indoor environment has been subjected to careful monitoring and measurement, with further industries springing up to quantify and control indoor air pollution. The Swedish Council for Building Research has, for instance, funded research "so that performance requirements may be specified for air quality" (Swedish Building Research, June 1994, p.7) and, in similar fashion, contributors to the Healthy Buildings '94 conference in Hungary aim to develop new knowledge in order to "build healthy buildings in place of sick ones" and to "convert sick buildings into healthy ones" (International Council for Building Research and International Society for Indoor Air Quality and Climate, 1994). Debate rages around the causes and circumstances of building sickness but the underlying fear is that in playing God with the indoor weather, in conditioning the air, in concocting new materials and in developing ever more precise climatic controls, building technologists have unleashed further environmental hazards, endangering the health of the very people they seek to protect.

Countering these suspicions, the air conditioning industry has drawn upon research which suggests that "the best defense against SBS (Sick Building Syndrome) could be a combination of an internal temperature of around 20°C, a relative humidity exceeding 20 per cent and a minimum fresh air flow of 10 litres per second per person", going on to argue that "the most effective, indeed possibly the only way of achieving these conditions would be by employing some form of air conditioning" (Jackson, 1994, p.30). In this spiralling of technological solutions further technology is proposed to counter risks and threats which appear to be related to existing methods of heating and cooling.

It is a difficult situation. Paradoxically, efforts to minimize damaging emissions of CO_2 and to reduce energy consumption by increasing insulation, by tightening the building envelope and by improving the technologies of energy management, have exacerbated fears and uncertainties about the reliability and quality of the indoor climate. As faith in the sciences of indoor climate control wanes, buildings take on a doubly monstrous role, threatenening both our environments, inside and out.

Social organization and the two environments

We must now confront some rather contradictory ideas. Buildings do keep the elements at bay and they do allow people to live and work in otherwise uninhabitable parts of the world. More than that, they can be designed to provide stable, comfortable and utterly predictable indoor climates. However, contentment with these indoor weather systems is increasingly tempered by anxiety about their effects on our health and welfare and by a growing recongition of their contribution to global destruction.

These confusions and uncertainties have immediate practical consequences for contemporary construction practice. The belief that buildings should respond to local climatic, even micro climatic, conditions is undoubtedly gaining ground. There is a tangible shift towards naturally ventilated buildings in place of the familiar high-tech capsules of the last twenty years. Systems of computer controlled natural ventilation are the latest invention, representing a glorious mixture of traditions in which the most up to date technologies are deployed to determine the opening and closing of windows. More radical advocates of "bioclimatic" architecture propose a return to traditional forms of building design, arguing for healthy structures, made from local, natural materials, each attuned to the special characteristics of its immediate environment.

But to what degree can we return to nature? Although the bioclimatic route appears to promise a safer indoor environment, one in which there are fewer sick buildings, in which there are no new potentially hazardous materials, and in which the boundary between inside and out is carefully, sensitively, managed, we are, it seems, propelled toward the design and use of ever more exclusive climatic fortresses. To understand why this is so we need to look again at the range of interests involved in using, making and maintaining buildings.

Institutional practices and cultural expectations have been transformed by, through, and with changing building technologies. Factors which started life as a set of engineering conventions, the set-point, the number of air changes per hour, the comfort zone, have become the norm for both building users and building designers. Such technical standards have helped to create a common climatic culture, overwriting historic, sociocultural variations in the meaning of comfort. Moving back in with the cows for the winter is simply not an option. Current lifestyles presume and to a degree depend upon a standardized, well managed, indoor environment.

In any case, the national and international organization of construction conspires against the widespread development of climatically sensitive, site specific, building design. Imagine the position of a private sector house building company in Britain. Such builders generally construct a limited range of "pattern book" houses in order to take advantage of associated economies of scale. For these reasons it makes sense to build the same house type on a north facing site in Scotland as on a south facing plot in Surrey. In this commercial environment there is no point in adapting plans, case by case, to suit the climatic characteristics of each location. Given a standard building layout and a standard form of construction, all that is then required is a solidly reliable, solidly oversized heating system capable of meeting all the demands ever likely to be made of it.

Similar pressures apply within the international construction market albeit for different reasons. Patterns of international investment have, for example, fostered the development of an increasingly limited range of building types (King, 1990). The very familiarity of the now standard office block provides a comfortable sense of security for investors and occupiers alike. Once established in the global culture and once reinforced by use and reuse, conventional understandings of appropriate building form take on a life of their own. As a result, office buildings in Tokyo look very like office buildings in Dallas or Dusseldorf; an indication, perhaps that the nature of office work, along with other patterns of social activity, is itself converging. Fast food chains and hotels similarly aim for, and achieve, worldwide standardization, inside and out. This international uniformity has important implications for the ways in which indoor climates are produced. Unable to significantly alter the conventional appearance of the office building, hotel, or shopping mall, and locked into the use of materials like concrete, glass and steel, designers and investors are driven to construct more or less "the same" building in any number of different climates. The task of managing the relationship between indoor and outdoor environments then becomes a task for the mechanical engineers. In other words, the pressures and economics of the international construction industry, together with a globalization of architectural imagery, effectively limit opportunities to make significant, climate specific, modifications to building form. In this context there is little option but to rely upon the machinery of heating and ventilation to handle the air and to exclude heat or generate warmth as required.

For its part, tne heating and ventilation industry has a vested interest in fostering demand for the equipment needed to create and sustain a standardized indoor climate. This industry, worth some $40 billion a year in 1992, is reported to be "enjoying five percent annual growth". With a market share of roughly 40 per cent each, the USA and Japan are exporting definitions of comfort as well as the means to achieve it (Sturm, Lord and Wagner, 1992, p.48). Rybczinsky notes that "the democratization of comfort has been due to mass production and industrialization" (Rybczinsky, 1987, p220) and it is indeed a measure of the success of the heating and ventilating industry that the ideal indoor climate is so carefully reproduced, parameter by parameter and building by building, the world over. The optimal environment is, it seems, best achieved, in many cases *only* achieved, by the use of mechanical systems of air conditioning. Comfort, as defined and quantified through scientific research, turns out to be a rather unnatural state of affairs. Baker spells out the implications of this observation. There is no doubt, he says, "that the conventional application of fixed comfort standards, as described by ASHRAE (the American Society of Heating, Refrigerating and Air Conditioning Engineers), CIBSE (the Chartered Institute of Building Services Engineers) and other regulating institutions would preclude passive cooling as an alternative to air-conditioning and commit our future built environment to a lifetime of high energy use" (Baker, 1993, p.106). In other words we have come to rely upon such a precise definition of comfort that we have no option but to exclude the outdoor environment and inhabit a world of carefully manufactured weather. Any more "natural" strategy will be subject to variation and unpredictability and so fail to meet the comfort standards we and our engineers have come to expect.

To summarize, heating and ventilation industries, together with building scientists and engineers have specified and created a global indoor climate. At the same time, increasingly standardized buildings accommodate increasingly convergent patterns of social activity. In these circumstances the outside environment must be excluded as efficiently as possible if there is to be any chance of maintaining the desired conditions inside. There is no place here for local variation, for site specific complexities or for the subtle interaction of indoor and outdoor climate.

Standing between the inside and the outside environment, buildings provide a tangible illustration of changing beliefs about the relationship between the two worlds. Taking concern about environmental change to heart means acknowledging complex interrelations between the different environmental spaces we inhabit. Now viewed as a vulnerable as well as a potentially inhospitable domain, the outside world is managed cautiously. Because of a growing recognition of the fragility of the global environment, buildings are redesigned to use less energy and to make fewer demands on limited natural resources. Perceptions of the indoor environment have changed as well. Technologies of indoor environmental control have begun to frighten the very people they were designed to protect and alternative, more "natural" methods of environmental management are sought in response. However, these more variable strategies of climate control run counter to the interests of the international construction industry, to the interests of those involved in heating and ventilating, and even to the interests of building occupants who have come to expect an utterly predictable indoor environment. Indoor climatic conditions, initially established with the aid of energy intensive technologies, have taken root in the international culture. Indeed this global climate revolution has been so wide ranging and so effective, that many taken for granted features of everyday life now depend upon the maintenance and management of a uniformly acceptable indoor environment. There is no way back to nature.

Accordingly, discussion concentrates on the technologies of energy efficiency, on future energy demand, and on the environmental consequences of CO_2 emissions. The challenge, it seems, is one of identifying commercially viable, environmentally friendly ways of sustaining the artificially uniform conditions to which we have become accustomed.

What is missing is more serious questioning of the indoor environment itself. While there is no way back to nature the process of indoor climatic convergence is neither inherently necessary nor totally irreversible. As suggested here, developments within the indoor environment form part of a complex sociotechnical process in which machinery is sold, definitions of comfort shift, social practices change, lifestyles evolve, and designers make decisions. We clearly need a much better understanding of these critical developments if we are to explore ways of minimizing outdoor environmental damage. Yet the indoor climate is not even on the outdoor environmental agenda.

By defining the environment as "everything out there", environmentalists have overlooked important changes in the indoor environment. They have, in effect, failed to notice the world wide manufacture of a standard indoor climate despite the fact that this is a development of considerable global cultural, technological, economic, and wider environmental significance. So it is time to refocus the environmental debate. It is time to come inside and to recognize the degree to which events in the outside world relate to what what goes on indoors.

References

Anderson, K., Norlen, U., Fagerlund, I., Hogberg, H., Larsson, B. (1992), *"Domestic Indoor Climate in Sweden: Results of a Postal Questionnaire Survey"*, Indoor Air Quality '92 Conference, "Environment for People", pp.18-21 October 1992, San Francisco.

Baker, N. (1993), "Thermal Comfort Evaluation for Passive Cooling" in Foster, N., and Scheer, H., (eds), *Solar Energy in Architecture and Urban Planning*, H. Stephens.

Beck, U. (1992), *Risk Society*, Sage Publications, London.

Boardman, B. (1991), *Fuel Poverty: From cold homes to affordable warmth*, Bellhaven Press, London.

Building Design (25.6.1982), "Police call in council for U-foam hit family", p.5.

Building Design (30.4.1982), "UF Foam: more cases found every week", p.1.

Building Research Establishment (1989), *Thermal Insulation: Avoiding the Risks*, HMSO, London.

Energy in Buildings (May 1994), "Split air conditioners help beat rising temperatures and falling productivity", p.6.

Fanger, P. (1970), *Thermal Comfort, Analysis and Applications in Environmental Engineering*, McGraw Hill Book Company.

Flandrin, J. (1979), *Families in Former Times*, Cambridge University Press, Cambridge.

Indoor Air International, Respiratory Disease and Indoor Air Pollution, conference leaflet 26 August 1994, Budapest, Hungary.

International Council for Building Research and International Society for Indoor Air Quality and Climate, Healthy Buildings '94: CIB-ISIAQ-HAS Conference, Budapest, Hungary 22-25 August 1994, *Second announcement and call for papers*.

Jackson, A. (1994), "Is air conditioning a waste of energy?", *Energy in Buildings and Industry*, February 1994.

King, A. (1990), "Architecture, Capital and the Globalization of Culture", in Featherstone, M., (ed.), *Global Culture*, Sage, Theory Culture and Society, London.

Morrill, J. (1994), "When the Straight Line is Not the Fastest Route: Counterintuitive Approaches to Improving Energy Efficiency in Buildings", ACEE 1994 Summer Study on Energy in Buildings

Rubin, A. and Elder, J. (1980), *Building for People*, Washington, National Bureau of Standards Special Publication 474.

Rybczynski, W. (1987), *Home: A Short History of an Idea*, Penguin Books, Harmondsworth.

Shove, E. (1991), *Filling the Gap: The Social and Economic Structure of the Cavity Wall Insulation Industry*, Institute of Advanced Architectural Studies, University of York, York.

Siddiqui, A. (1990), "Air conditioning strategies and implications vis a vis social and environmental factors in the city of Jeddah, Saudi Arabia" Paper 1.8 in *International Symposium on Energy Moisture and Climate in Buildings*, CIB W67 Symposium 3-6 September 1990, Netherlands.

Sturm, R., Lord, D. and Wagner, L. (1992), *Seizing the moment: Global Opportunities for the U.S. Energy-Efficiency Industry*, International Institute for Energy Conservation, Washington D.C.

Swedish Building Research (June 1994), "Odour Research for Healthy Buildings", p.7.

Tyler, M. (1991), "Sick Buildings", *The Architectural Journal*, Vol 194, pp. 48-50.

Part II
NATIONAL AND INTERNATIONAL POLITICS OF THE ENVIRONMENT

1 Environmental concern in Britain 1919–1949: Diversity and continuity

Ian Coates

Introduction

Environmentalism is generally seen as a comparatively recent phenomenon, with most accounts of its development beginning in the 1960s and 1970s. Yet a historical perspective can reveal that many of the environmental concerns experienced today are hardly new, as can be demonstrated by examining reactions to similar issues in Britain between 1919 and 1949. This was a period of structural change which had a profound effect on the environment and attitudes towards it, eliciting responses from all sectors of society. Whether such responses could be described as environmentalism, as it is understood in its contemporary sense, is open to question since it would be misleading to project current worldviews onto actors operating within different historical circumstances. However, such a focus can allow a fresh interpretation and demonstrate striking parallels between environmental concern in the past and that of the present.

There have been a number of attempts to re-evaluate events in Britain between the wars by highlighting their environmental aspects, which, though of great value, have tended to focus on particular individuals, interest groups, and areas of concern.[1] With the exception of Lowe's work on the historical development of environmental groups, few accounts have sought to bring the separate strands together and consider the range of responses in a broader social context.[2] Whilst Lowe does not cover interwar Britain in great detail, his approach can be extended to consider the diversity of environmental concern in this period in terms of structural and environmental change, social movements and institutions. This can be illustrated by selected examples representative of different forms of environmental concern which can be differentiated in terms of the interests, discourses and activities of those involved. These include the traditionalist and nationalist forms of ruralism present

61

amongst sections of the political right; the experiential and holistic outlook of groups concerned with food, health and nature; the popular appreciation of the countryside represented by organizations like the Ramblers' Association; and the planning and conservationist approach of bodies like the Council for the Preservation of Rural England.

The typology suggested here is not meant to be definitive, but merely offers a means to organize the material under discussion. Categorization schemes are inevitably arbitrary and when situating relevant groups within this one, some could be placed within a number of categories, whilst others are more difficult to classify and fall between the suggested types. All that is intended is to indicate the range of responses involved and to show that environmental concern can take a number of forms as diverse groups experience, interpret, and construct the environment by way of their own outlook and interests. These particular examples have also been chosen because in some form they can be seen to have perpetuated themselves through to the present day, having a lasting effect on the shape and use of the environment, as well as on perceptions of it. Though a global worldview and the specific problems it confronts distinguishes contemporary environmentalism from that of the past, continuities are also apparent in which today's concerns can be seen as part of an ongoing process. Parallels can also be drawn between the variety of forms of environmental concern in Britain between the wars and in the present.

Rural traditionalism and English nationalism

In *Ecology in the 20th Century*, Anna Bramwell implied that political ecology as an idea originated on the right of the political spectrum and displayed the influence of nationalist and racial elements, citing a number of examples from interwar Britain.[3] As will be seen, it was certainly the case that some individuals and groups on the fringes of the right displayed an awareness of environmental issues, but this tended to be grafted onto a more established conservative discourse of rural traditionalism and English nationalism, which in its dominant forms had little to do with protecting the environment.

There is a long history of groups on the right portraying themselves as the guardians of a rural tradition in which images of England's green and pleasant land were employed to legitimize hierarchical social relationships and foster patriotic appeals to a British national identity. Elements of such a discourse can be found in the speeches of Stanley Baldwin who was leader of the Conservative Party from 1923 to 1937. These commonly drew on idealized images of Britain's rural heritage in a language which sought to generate a popular appeal by connecting the British national character to the "eternal traditions" of country life.[4] This could be seen as a response to a growing urban working class, constituting a defensive reaction to the social changes challenging old institutions and values.[5]

Rural nostalgia and arguments in favour of agricultural self sufficiency could also be associated with notions that "country stock" was somehow physically, morally, and even racially more healthy than urban populations, partly reflecting anxieties over the perceived growth of immigrant communities (particularly Jewish ones). Such an explicitly racial rhetoric was used by organizations like Oswald Mosley's British Union of Fascists, who contrasted urban degeneracy and "cosmopolitanism" with the "virile country stock" of the "yeomen of England".[6] More marginal groups such as the Britons also saw rural society in terms of racial identity. As well as publishing antisemitic tracts, the Britons distributed books on the soil, health and organic farming, and blamed a conspiracy of Jews, Communists and Freemasons for pollution, pesticides and artificial fertilizers, as part of a plot to undermine the bodily purity of the British nation.[7]

A similar group on the fringes of the far right were the English Array, founded by Viscount Lymington and Rolph Gardiner, both staunch supporters of Nazi Germany.[8] The writings and activities of these two figures are of particular interest because as farmers and landowners they both took a practical interest in the techniques of organic farming and in promoting these ideas to a wider audience. What was striking about Lymington and Gardiner was that alongside their extreme racial politics they showed a considerable ecological awareness in their analysis of the interrelationship of social and natural systems. They saw land management as part of an integrated whole involving forestry, wildlife conservation and mixed farming using organic methods as well as the development of local industry to help sustain rural communities. Present in their work was a grasp of issues such as the dangers of monoculture and factory farming, soil erosion, declining land fertility, over abstraction of ground water, the need to utilize sewage wastes as fertilizers and the importance of preserving hedgerows and woodlands to encourage natural predators able to control pests.

Lymington and Gardiner were not the originators of these ideas which were largely synthesized from the work of scientists like Jacks and Whyte, R.G. Stapledon, Albert Howard, and Ehrenfried Pfieffer.[9] Also, given their political outlook, these ideas were interpreted in a particular way. Lymington argued that the issue of self sufficiency in food was one of national and racial survival, requiring the utilization of every available piece of land. He also wanted the restoration of the central role played by landowners to provide capital, resources and leadership for a reconstituted peasantry. Diversified rural production, using organic methods, was to result in healthy food, healthy bodies, healthy minds, and a healthy nation.[10] Gardiner, expecting the imminent collapse of industrial society, saw a need to build centres able to preserve the values of the past from which could emerge a revitalized European rural civilisation. In an attempt to bring such a centre into being Gardiner restored a group of farms at Springhead in Dorset, using forestry and organic techniques, as well as encouraging local craft industries and promoting cultural links between North European countries.[11]

Following a conference organized by Lymington to discuss organic farming methods, some of those attending, along with contributors to H.J. Massingham's *England and the Farmer*, established Kinship in Husbandry.[12] This was a discussion group which held occasional meetings between 1941 and 1950, attended by between 12 and 20 writers and farmers who could best be described as High Tories. Their publications deplored the consequences of industrialism and looked to an idealized traditional rural society as a model for a less exploitative way of life. A return to the land and the ownership of private property was seen as a counter to the destructive effects of industrial farming methods, state intervention and the soullessness of modern life. There was little sign that other members of the group shared Lymington's and Gardiner's pro Nazi or racist beliefs. Certainly Massingham was strongly anti Nazi and thought that racial theories were nonsense.[13]

Bramwell places a great deal of emphasis on Lymington and Gardiner and their links with Nazi Germany which she appears to have seen as the first green state, suggesting that they were a principal influence on the organic farming movement in Britain, by way of their early membership of the Soil Association. However, this organization was not formed on their initiative and they did not subsequently play a very active role in its affairs.[14] As will be seen, the Soil Association, though sharing similar concerns over the consequences of modern agriculture, did not see solutions to this in terms of a return to a traditional society. The Soil Association's journal, *Mother Earth*, certainly showed no evidence of racial or nationalist concerns, and took no overt political stance, and was exclusively concerned with giving practical advice and presenting a scientific case for organic farming.

Bramwell's work exaggerates the importance of the historical association between the right and some forms of environmental concern. Though undoubtedly present, this approach involved relatively few people and it remained marginal in effect. As well as being based on a highly selective interpretation of the evidence, Bramwell's account is limited by a narrow definition of what constitutes political ecology, which in practice seems to amount to little more than national self sufficiency combined with environmental conservation. If broader social factors are considered it can be seen that rural concern in this period was by no means exclusive to the right. All political parties expressed the desire to prevent rural decline, though this was rarely translated into concrete action to remedy the situation. Criticism of the negative effects of modern industrial society was just as developed amongst sections of the left, following in the radical tradition of William Morris and Edward Carpenter. This alternative current could be detected in the writings of a number of prominent socialists of the period, as well as in the programme of the Guild Socialists or in the calls for land nationalization and rural industrial diversification made by Montague Fordham's Rural Reconstruction Association.[15]

Experiential holism: the Soil Association

Concern over the countryside and human relationships with nature could also take other forms which are far less easy to classify in political terms, such as the pursuit of alternative lifestyles involving a return to the land to live a simple life through gardening or smallholding, arts and crafts, health food diets, vegetarianism, or naturism. This category might best be described as experiential holism. One organization that represents aspects of this kind of response was the Soil Association.

The Soil Association had its origins in the experimental programme undertaken by Eve Balfour. She established a research trust to compare organic and nonorganic farming methods in terms of their effects on the soil and the quality of the food produced. The experiment was described in her book *The Living Soil*, which also presented a summary of recent research on the link between soil fertility, food quality, and human health.[16] Balfour, a critic of modern farming methods, sought scientific validation to support alternative approaches. Though her quest for wholeness undoubtedly drew on spiritual and vitalist currents, she also looked to the new science of ecology to outline the wider consequences of attempting the conquest of nature through industrial and chemical farming. Her discussion of these issues was also set in the context of contemporary debates concerning postwar reconstruction and the need to plan and build a better society. Though not taking a specific political stance, Balfour favoured a democratic and decentralized approach, arguing that private ownership of land led to over exploitation and that some form of public ownership might be a better option.

Following the positive response to her book, the initial meeting of the Soil Association in 1945 was attended by over one hundred people and by 1947 it had a membership of more than one thousand.[17] The aims of the Soil Association can be summarized as being: to bring together all those working towards a fuller understanding of the vital relationship between soil, plants, animals and humans; to initiate, coordinate and assist research in this field; and to collect and distribute information to create a body of informed public opinion.[18] Though successful in its role as an advisory and coordinating body for the organic farming movement, the Soil Association's experimental programme was always short of the kind of funds necessary to run a long term comparative trial. This proved a drain on the Soil Association's meagre resources and after a series of financial crises the research had to be abandoned. A related problem was that despite the involvement of a number of reputable scientists, the Soil Association's research programme was dismissed as the work of cranks by a scientific and policy community committed to a chemical and energy intensive agriculture.

British agriculture and rural decline

These concerns over farming methods, food production and rural society need to be set in the context of the crisis facing British agriculture. British farming had been in decline since the 1870s, partly as a result of competition from cheap food imports. Policies of agricultural support during the Great War had allowed a brief revival, creating a new wealthy class of farmers. At the end of the war up to a quarter of all rural property changed ownership with the break up of many large estates. A large proportion of this land was bought up by farmers themselves, but prices were high which left a future legacy of heavy mortgages. The boom was brief and within a few years falling land and food prices, combined with higher costs, led to reductions in wages and cuts in the workforce, driving labour from the land. Meanwhile, as rents fell and taxes rose, farmers faced increasing difficulties, and with large areas of land being taken out of cultivation, much of the countryside showed visible signs of neglect.[19]

This trend was reversed by government requirements to increase domestic food production with the onset of World War II. This policy continued given the continuing food shortages and rationing of the postwar period. The government had finally come to recognize the importance of assisting domestic agriculture with the Report of the Scott Committee in 1942, many of whose recommendations were embodied in the 1947 Agriculture Act. This guaranteed prices to farmers whilst providing them with subsidies to increase efficiency. Thus the Soil Association's argument for an alternative approach was marginalized in the face of the needs of government and consumers for cheap and plentiful food supplies, and the vested interests of the chemical companies and those favouring large scale agricultural units. It was only many years later when the economic and environmental consequences of subsidizing an energy intensive, industrialized agriculture became apparent that the case for organic agriculture would come to be seriously re-examined.

The countryside between the wars

These approaches to the environmental issues of the day reflected a more widespread concern over the countryside and rural life in the interwar period. This was not solely due to the rural decline brought about by the crisis in British agriculture. At the same time urban areas were expanding, increasing by 26 per cent, with four million houses being built within two decades, mostly to enable the middle classes to move away from old and decaying housing stock in the cities to new homes in the suburbs.[20] Improved transportation, through by cheap rail travel and growing car ownership, contributed to fears that ribbon development would destroy the distinction between town and country. Concern was also voiced that increased motoring for leisure would shatter the peace of the countryside and lead to the building of petrol stations, roadhouses and teashops to cater for this new pastime. An associated phenomenon was the increasing number of popular books, magazines and guides, celebrating the English countryside and encouraging people to go out and discover it for themselves.[21] These tended to describe rural life and the landscape in rather idealized terms, largely ignoring the conditions of those who actually lived and worked there.

Another development was the use of rural imagery in order to advertise. Shell's campaign was the most notable, especially as the oil companies were seen as contributing to the destruction of the peace and quiet of the countryside. To counteract this negative image, in 1930 Shell announced that they would no longer advertise on roadside hoardings, instead commissioning artists to design posters of rural scenes to be displayed on the sides of lorries with slogans such as, "See Britain First on Shell". In 1934 the first Shell Guides to Britain appeared. As Shell themselves put it some years later, "The primary objective of all this was strictly business - to encourage motorists to explore the highways and byways, using Shell petrol and lubricants".[22] Such commercialization amounted to an objectification and commodification of the countryside. Since it was no longer seen as a place where people lived and worked, it could become an object to be appreciated, and because it was desirable, used to sell products.

With the decline of the rural economy, and as fewer people had direct contact with the land, the myth of Arcadia seemed to increase in power. Yet this myth underwent changes in emphasis for different social groups. The middle class move to the suburbs could be seen as not only an escape from the squalor of urban life, but also from its restrictions, to a place where they could assert their own identity in a way that reflected their changing social position. In the suburbs they could experience more open space, a more affluent lifestyle, and engage in a range of leisure pursuits. Regarding their relationship with the countryside this was not particularly understood in terms of a retreat into the past, but was rather experienced in quite a modern way, to be enjoyed as a place of recreation where they could pursue activities seen as conferring benefits of physical and psychological health. This manifested itself in various forms of popular appreciation of the countryside.

Popular appreciation of the countryside: the Ramblers' Federations

Estimates of the numbers of people involved in outdoor activities in the 1930s have been placed as high as 500,000.[23] Walking was one of the most popular outdoor pastimes, especially with young people, from both working class and middle class backgrounds. In the interwar period hiking became associated with ideas of health, fitness and social permissiveness. For those experiencing the enforced leisure of unemployment it also had the advantage of being a cheap means of escaping from the bleakness of urban life. Bus and railway companies often organized special excursions and granted concessionary fares to ramblers.

The first rambling clubs had been established in the 1880s and 1890s, a number of which were associated with the *Clarion*, a socialist newspaper. The Clarion Field Clubs were sociable and family oriented, aiming to cultivate an appreciation of the countryside by bringing, "the town dweller more frequently into contact with the beauty of nature, to help forward the idea of the simple life, plain living and high thinking", the idea being that creating better people would help build a better society.[24] The same ethos informed groups like the Sheffield Clarion Ramblers, founded in 1900, who were to significantly influence the direction of the rambling movement.

By the 1920s the individual rambling clubs were beginning to form local Ramblers' Federations to campaign more effectively on such issues as gaining free access to moorland, to preserve rights of way, to obtain cheap rail fares and to protect the countryside. A National Council of Ramblers' Federations was formed in 1931 which by 1935 became the

Ramblers' Association with over 50,000 members.[25] The decision to form their own national association partly resulted from the perception that though the major amenity groups, like the National Trust and the Council for the Preservation of Rural England, supported increased access, at the same time they gave priority to the interests of large landowners.

The lack of free access to private land remained the key issue for ramblers. This was less of a problem in the South where public footpaths were well established and walkers on downland and pasture were not seen as a threat. In the North of England where rambling was much more a mass working class movement, the upland areas near the major cities were predominantly used for grouse shooting. Claiming that free access would cause damage and loss of income, large landowners employed keepers, generally armed with clubs or guns and accompanied by dogs, to keep walkers out. Even some council owned land was subject to a no access policy, either on the grounds of possible contamination of water catchment areas, or because the councils themselves were letting the land for private shooting.[26]

The Ramblers' Federations responded with a series of rallies and demonstrations, with the Manchester and Sheffield Federations being amongst the most militant. Their annual demonstration at Winnats Pass in Derbyshire attracted 10,000 protesters in 1932. Also in 1932 the British Workers' Sports Federation organized a mass trespass near Kinder Scout, a peak between Manchester and Sheffield, where only 764 out of 84,000 acres of moorland were open to the public. About 500 protesters were met by 60 keepers, one of whom was knocked unconscious. Arrests followed and five protesters received prison sentences.[27] The Ramblers' Federations were not actively involved, preferring to stay within the law and work through negotiation, though some did organize mass trespasses themselves. Such events were certainly a symptom of the resentment felt by ramblers on this issue. An Access Bill eventually became law in 1939, but ramblers regarded this as a landowners' charter. It was only in 1949 with the Access to the Countryside Act and the National Parks Act, in which the Ramblers' Association had taken an influential consultative role, that wider rights of access were granted.

Whereas on one level this was a conflict about access for recreation, it also raised wider issues of property rights, land use and environmental impacts. The owners were arguing that free access would lead to environmental damage and loss of income, but the use of moorland areas for private shooting meant that other forms of land use were not considered. As these moors were exclusively managed to protect game this led to the systematic eradication of all predators, including rare species, with all the undesirable ecological consequences

69

resulting from such a policy. People were also seen as a threat to commercial interests, even though it was shown that grouse were not affected by public access. The ramblers' attack on property rights could be seen as not merely to guarantee legal access, but also to preserve the character of the landscape, its flora and fauna, to be enjoyed for their own sake.

Corporate conservationism: the Council for the Preservation of Rural England

The question of access to the countryside had also involved the Council for the Preservation of Rural England. In 1928, along with the representatives of other interested groups, they had organized a Countryside Conference where the Federations of Rambling Clubs had demanded an Access to Mountains Bill. Though the CPRE was in favour of increased access it wanted this to be controlled, and was certainly opposed to it being, "without let or hindrance". This could be explained by the fact that large landowners were one of the interest groups represented by the CPRE, which drew its support from institutions which were very much part of the establishment.

There were already a number of amenity and conservation groups in Britain, dating back to 1865 with the founding of the Commons Preservation Society. This was followed by, amongst others, the Society for the Protection of Ancient Buildings, the Society for the Protection of Birds, the Society for Checking Abuses in Public Advertising, the National Trust and the Society for the Promotion of Nature Reserves.[28] Though campaigning on behalf of the public interest, these groups tended to stick to their specialist area and worked to bring about the changes they desired through appeals to property interests and political elites. The National Trust epitomized this approach, itself becoming a large landowner and forming a close relationship with government, which almost regarded it as a public agency. However, this also limited the scope of its activities.

The CPRE was brought into being in order to undertake a role not being met by existing amenity groups. The initiative to found the CPRE had come from Patrick Abercrombie, a Professor of Town Planning who wished to find a balance between the needs of modern life and aesthetic considerations. Abercrombie thought that through planning it would be possible to exploit the countryside without destroying its essential qualities.[29] The inaugural meeting of the CPRE in 1926 was attended by the representatives of twenty organizations including established amenity groups like the National Trust, along with other interested parties such as the Royal Institute of British Architects, the Town

Planning Institute, the Urban, County and Rural Councils Associations, the Central Landowners Association, the Country Gentlemens' Association, and even the AA and the RAC.[30] Within ten years the CPRE represented 42 constituent bodies, 140 affiliated bodies, 28 local county branches and a large associate membership, all contributing towards administration costs of £5,000 per year for a headquarters and permanent staff.

The CPRE stated its aims as being to protect rural scenery and amenities, to provide advice and inform policy on these issues, and to educate public opinion. Other activities undertaken by the CPRE included the coordination of the work of different amenity groups, with whom they often formed special joint committees on particular issues, commissioning surveys and reports, and the organizing of exhibitions and lectures around the country. The CPRE had considerable influence, overseeing planning schemes at national and local levels, as well as formulating and promoting legislation. In these roles it was regularly consulted by central and local government and various public bodies. Though the CPRE sought to work with the government they were highly critical of a lack of policy and the absence of coordination between government departments.

An early aim was to limit urban sprawl and ribbon development, which the CPRE did to some extent through its input into the 1932 Town Planning Act, the 1935 Ribbon Development Act and the 1938 London Green Belt Act. Another initiative began in 1929 when the CPRE submitted a memorandum to the Prime Minister on the need to create National Parks in Britain. A commission was established and a report published, but when the government failed to implement it, the CPRE formed the Standing Committee on National Parks to campaign for their establishment and to generally provide more public access to the countryside. This objective was finally achieved with the creation of the National Parks in 1949. However, the CPRE was also concerned to minimize the impact of public access, which they sought to do through an education programme promoting a country courtesy code and by the provision of Countryside Wardens wearing the CPRE badge.

Though the CPRE was undoubtedly made up of elite and establishment groups, including landed interests, it did take on an active campaigning role on a number of issues and took a critical stance to many government bodies. It cannot be seen as a backward looking movement concerned simply with preserving the past. The CPRE was arguing for a planned approach that would achieve a balance between the needs of modern life and conserving the aesthetic appearance of the landscape. Often the conception of what constituted an aesthetically pleasing landscape was surprisingly modernist in outlook. For example Abercrombie was not against motorways as long as they were sited with care.[31] They saw their role as being, "to promote suitable and harmonious development and to encourage rational enjoyment of rural areas by urban dwellers", allowing the physical and mental improvement of the citizen.[32]

Conclusion: from the past to the present

The intention of this necessarily brief and selective survey has been to indicate just some of the varied forms of environmental concern in Britain between 1919 and 1949. Given the profusion of relevant groups which arose during this period, not to mention those already established, it has not been possible to include them all. For example there has been little discussion of organizations concerned with nature conservation or of state responses to environmental issues, whether by legislation or through the creation of agencies like the Forestry Commission. Rather, the aim has been to show how concern over the environment arose in response to economic, technological and social factors, such as the decline of traditional agriculture, rural communities and populations and the corresponding expansion of urban areas. Increasing social mobility and leisure can also be seen as having contributed to further impacts on the countryside as urban populations found it increasingly desirable as a place of habitation and recreation.

These changes affected groups from all sectors of society, each with their own set of interests, perceptions and activities, which led to the development of different responses and programmes concerning the use of the countryside and impacts upon it. Though the countryside was often an arena of conflict between different interests, areas of concern could overlap, as could personnel. Thus, there were often common interests and shared objectives between different groups, such as the campaign to establish National Parks which involved the Ramblers' Federations and the CPRE. Just as the contemporary environmental movement includes a diversity of views, strategies and goals, alongside areas of general agreement, the same could be said of environmental concern between 1919 and 1949.

This allows the question to be raised as to whether the concern shown over the countryside in the interwar period can be seen as a manifestation of environmentalism in the sense that this is understood today. If environmentalism is taken to be an active concern over human impacts on the environment and their long term consequences for natural systems and human societies, it can be seen that there are many ways in which aspects of earlier concerns conform to this definition. Yet contemporary environmentalism can certainly be distinguished from past forms on a number of grounds. These include the particular economic and structural conditions of modern urban industrial society that gave rise to it; the specific problems it confronts; the increased awareness of threats to human and nonhuman environments; its global perspective; and its ability to become a political force seeking a radical transformation of society. However, many of the local issues faced today are not new and there are several ways in which the contemporary environmental movement could be seen as reflecting earlier concerns, or even having been shaped by them.

In considering the forms of environmental concern prevalent in the interwar period, it can be seen that they have all perpetuated themselves to the present day, whether institutionally, or as a set of attitudes and discourses. This is perhaps less the case with rural traditionalism and English nationalism, though continuities can certainly be drawn between the racial nationalism articulated by groups like the Britons and English Array, and the ideology of the National Front, whose attempt to adopt environmental concerns between 1980 and 1990 took a similar form.[33] Nevertheless, it should be remembered that these groups were marginal then, just as the National Front is marginal now.

Attitudes similar to those advocated by Kinship in Husbandry can also be found on the New Right, as in Roger Scruton's argument that private ownership of land will preserve the landscape, and that it is in the traditions and rituals of rural life that English national culture is best preserved. For example, he characterizes foxhunting as an ecological institution and contends that rural familiarity with animal breeding teaches that inheritance matters.[34] Elements of this kind of discourse are similarly present in Conservative attitudes towards the environment or in institutions like the National Trust who perpetuate images of traditional England through their emphasis on preserving great country houses and large estates. So, such appeals can be implicitly present in the discourse of conservation of Britain's national heritage. Yet such a backward looking preservationism is a long way from the predominant forms of contemporary environmentalism. Few environmentalists today seek a return to an idealized past and there is little scope for an exclusively national response to environmental problems in the age of acid rain, ozone depletion and global warming.

More significant parallels can be found for the other forms of environmental concern under discussion. Not only are many of the arguments, attitudes, practices and responses very familiar, but these groups have also shown an ability to build institutions which continue to affect how the environment is understood and used today. In this they have acted as agents for change, defining problems and working

73

towards solutions, much as comparable organizations do today. The Soil Association would seem to have directly influenced the newly emerging environmental movement of the early 1970s, in terms of its holistic philosophy, its emphasis on lifestyle and in its attempts to find sustainable alternatives to intensive agriculture.[35] However, the Soil Association has also had a significant effect in its own right through its ability to provide expertise, standards and regulation for a growing organic farming movement, to the extent that it is now regularly consulted by government agencies on issues such as environmental labelling and sustainable agriculture.[36]

Grassroots organizations such as the Ramblers' Association, mounting mass campaigns and demonstrations, as well as undertaking consultative or negotiating roles, show many similarities with the tactics of current environmental campaigning groups. They are successful in opening up large areas of countryside and are still active in challenging restrictions to access and protecting the rights of walkers, as well as campaigning on broader environmental issues such as their recent opposition to the privatization of the Forestry Commission. In many ways their mass campaigns in the 1920s and 1930s could be seen as a prototype for later mass actions by environmental groups.

Similarly the professional approach of a tightly organized pressure group like the CPRE, able to conduct public education and media campaigns, as well as formulating policy and influencing legislation, also shows parallels with a number of environmental lobbying groups today. With its central organization, local branches and large membership, the CPRE has remained an effective environmental lobbyist, having a considerable influence on planning legislation. Despite its close relationship with government, it has, if anything, become more outspoken over time. Recent campaigns include its challenge to the expansion of open caste mining operations and its opposition to an accelerated road building programme. Yet there are also limitations to the strategy of lobbying government and working within the planning system in that it fails to challenge adequately the underlying economic and political causes of environmental problems, thus leaving scope for other more radical approaches.

Though these groups, representing different forms of environmental concern, were shaped by the time and conditions within which they arose, it is possible to see them as being concerned with a number of similar problems to those facing the environmental movement today. These examples also reveal the range of activities involved as groups with different interests, experiences and intentions concerning the environment constructed different sets of responses. It is as though when new interests and activities brought people into contact with a countryside experiencing structural change, this not only allowed it to be used and seen in new ways, it allowed human impacts upon it to be understood in new ways. Looking back at the diversity of forms of environmental concern in the interwar period, it can be seen that this is paralleled by a comparable diversity in the present. It can also be argued that in many ways the contemporary environmental movement did not just suddenly arise in the 1960s, but rather that it is part of an ongoing process.

Notes and references

1.   Regarding Britain between the Wars there are a number of accounts dealing with the development of conservationism and its impact on policy. See for example, Sheail, J. (1976), *Nature in Trust*, Blackie, Glasgow & London and Sheail, J.(1981), *Rural Conservation in Inter-War Britain*, Clarendon Press, Oxford. On organic farming see Conford, P. (1981), *The Organic Tradition*, Green Books, Hartland. For an overview of the subject of green history see Chase,M. (1992), "Can History be Green?", *Rural History*, Vol 3, No 2, pp. 243-251.

2.   Lowe, P. "Values and Institutions in the History of British Nature Conservation", in Warren, A. and Goldsmith, F. (eds) (1983), *Conservation in Perspective*, John Wiley & Sons, Chichester, pp. 329-352; Lowe, P. "The Rural Idyll Defended: from Preservation to Conservation", in Mingay, G.E. (ed) (1989), *The Rural Idyll*, Routledge, London, pp. 113-139; and Lowe, P. and Goyder, J. (1983), *Environmental Groups in Politics*, Allen & Unwin, London.

3.   Bramwell, A.(1989), *Ecology in the 20th Century*, Yale University Press, New Haven.

4.   Baldwin, S. (1937), *On England*, Penguin, Harmondsworth, p. 16. See also, Schwarz, B.,"Conservatism", in Donald, J. and Hall, S. (eds) (1986), *Politics & Ideology*, Open University Press, Milton Keynes, pp. 154-186.

5.   See Howkins, A.,"The Discovery of Rural England", in Colls, R. and Dodd, P. (eds) (1986), *Englishness and National Culture 1880-1920*, Croom Helm, Beckenham, pp. 62-88, and Wiener, M. (1981), *English Culture and*

the Decline of the Industrial Spirit 1850-1980, Cambridge University Press, Cambridge.

6. Mosley, O. (1936), *Tomorrow We Live*, Greater Britain Publications, London, p. 46.

7. See Thayer, G. (1965), *The British Political Fringe*, Anthony Blond, London, chapter 6.

8. See Griffiths, R. (1983), *Fellow Travellers of the Right*, Oxford University Press, Oxford.

9. Some of these figures shared similar political prejudices to those of Lymington and Gardiner. Jacks and Whyte of the Imperial Bureau of Soil Science at Rothamsted, had published (1939), *The Rape of the Earth*, Faber, London, a world survey of soil erosion which, though highlighting the extent of the problem, was also full of nationalist, imperialist, and racist assumptions. Sir George Stapledon, Director of the Welsh Plant Breeding Station, an eminent plant geneticist who had studied with the ecologist A.G. Tansley, also subscribed to racial ideas in that he believed that country stock was genetically more sound than town stock, as in his (1935), *The Land Now and Tomorrow*, Faber, London, p. 231. Stapledon was not pro Nazi and could better be described as a High Tory. It is not apparent that Howard, the inventor of modern composting methods, and Pfieffer, an advocate of Rudolph Steiner's biodynamic farming, shared the political views of Lymington and Gardiner.

10. Viscount Lymington, "The Policy of Husbandry", in Massingham, H.J. (ed.) (1941), *England and the Farmer*, Batsford, London, pp. 12-31. Also see Lymington (1932), *Horn, Hoof and Corn*, Faber, London, and Lymington (1938), *Famine in England*, Right Book Club, London.

11. Gardiner, R. in Best, A. (ed.) (1972), *Water Springing From the Ground*, Springhead Publishing, Springhead.

12. Massingham, H.J. (ed.) (1941), *England and the Farmer*, Batsford, London.

13. See Blunden, E. (ed.) (1943), *Return to Husbandry*, Dent, London; Massingham, H.J. (ed.) (1945), *The Natural Order: Essays in the Return to Husbandry*, Dent, London; Massingham, H.J. (1941), *Remembrance*, Batsford, London; The Earl of Portsmouth (formerly Viscount Lymington) (1965), *A Knot of Roots*, Bles, London; and Gardiner, R., *Water Springing From the Ground*, op. cit.

14. Though elected onto the Soil Association's Council, the minutes, as published in *Mother Earth*, invariably record both Lymington and Gardiner as absent. They also did not significantly contribute to the Soil Association's journal.

15. On the inaugural meeting of the Rural Reconstruction Association see *The Times*, 25 March 1926. On socialist rural concern see Wiener, op. cit., pp. 118-126.

16. Lady Eve Balfour (1941), *The Living Soil*, Faber, London.

17. *Mother Earth*, No 4, 1947.

18. *Mother Earth*, No 1, 1946.

19. See Mingay, G.E. (1990), *A Social History of the English Countryside*, Routledge, London and Howkins, A. (1991), *Reshaping Rural England: A Social History 1850-1925*, Harper Collins, London.

20. See Beddoe, D. (1989), *Back to Home and Duty: Women Between the Wars 1918-1939*, Pandora, London, pp. 95-96.

21. These included Morton, H.V. (1927), *In Search of England*, Methuen, London, magazines like *The Countryman*, and series of titles such as the "Open Air Series" published by Dent from 1932, and Batsford's "British Heritage and Face of England Series".

22. Shell UK (1988), *Good Business*, Shell UK, London, p.2.

23. See Lowerson, J., "Battles for the Countryside", in Gloversmith, F. (ed.) (1980), *Class, Culture & Social Change: A New View of the 1930s*, Harvester, Brighton, pp. 258-280.

24. Prynn,D. (1976), "The Clarion Clubs, Rambling and the Holiday Associations in Britain Since the 1890s", *Journal of Contemporary History*, No 11, pp.65-77.

25. Stephenson, T. (1989), *Forbidden Land*, Manchester University Press, Manchester.

26. Sheffield Corporation banned their own ratepayers from council owned land until 1936. See Hill, H. (1980), *Freedom to Roam: The Struggle for Access to Britain's Moors and Mountains*, Moorland Publishing, Ashbourne.

27. See accounts in Hill ibid., Lowerson op. cit., and Step henson op. cit.

28. Lowe, P., "The Rural Idyll Defended: from Preservation to Conservation", op. cit.

29. Abercrombie, P. (1937), "Country Planning", in Clough Williams Ellis (ed.) (1937), *Britain and the Beast*, Dent, London, pp.133-140.

30. *The Times*, 8 December 1926, p.13.

31. Abercrombie, P. (1926), *The Preservation of Rural England*, University Press of Liverpool and Hodder & Stoughton, London, p.25.

32. Sheail, J., *Rural Conservation in Inter-War Britain*, op. cit., p.64.

33. See Coates, I. , "A Cuckoo in the Nest: The National Front and Green Ideology", in Holder, J. et al. (eds) (1993), *Perspectives on the Environment*, Avebury, Aldershot, pp.13-28.

34. Scruton,R., "A Green and Pleasant Land", BBC 2, 22 October, 1991.

35. There was a direct connection between the Soil Association and the contemporary environmental movement through M. Allaby and R. Waller who were simultaneously editors of *Living Earth* (formerly *Mother Earth*) and *The Ecologist* at the time of the publication of the influential *Blueprint for Survival* in 1972. Fritz Schumacher, author of *Small is Beautiful* was President of the Soil Association at this time.

36. Though organic farming only amounts to one per cent of British agriculture the Soil Association certify seventy per cent of this acreage.

2 The Ukrainian green movement: Nationalist or internationalist?

Åse Berit Grødeland

Introduction

In a paper presented at a conference on the Soviet Environment at the Hebrew University of Jerusalem in early January 1990, Marshall Goldman argued that whereas the Greens in the West are "internationalist", the Green Movements emerging in the Soviet Union were separatist (nationalist) in character: "Inside the Soviet Union, environmentalists tend to be separatists...In contrast, the environmental cause in most of the rest of the world tends to be anti-nationalistic, almost "one world" in outlook."[1] Representatives of green circles in the West have also expressed views similar to that of Goldman. The Greens in the former Soviet Union, though, are not at all happy with being referred to as "nationalists" and have on numerous occasions stated their commitment to the principle of "thinking globally and acting locally".

In an interview with the Ukrainian weekly *Molod Ukrainy* in February 1992 Vitaili Kononov, currently the leader of the Green Party of Ukraine, voiced his dissatisfaction with this situation, stating that "the Greens in the West do not perceive of us in the way we would have liked them to. We are being accused of nationalism".[2]

In this article, I hope to demonstrate that the situation is more complex than argued by Goldman and, although the national question is an issue on which the Green Movements have been forced to make a stand, that the motivation behind what I will refer to as "national sentiments" amongst the Greens is qualitatively different from nationalism in its negative sense of territorial expansionism and/or the primacy of one ethnic group in a given territory. I will argue that "national sentiments" have been adopted by the Greens only as far as they are required to stabilize/improve the state of the environment and that, like their sister movements/parties in the West, the former Soviet Greens too are deeply committed to the principles of the international green movement.

I will focus on Ukraine and examine the attitudes of the Green Movement (Zelenyj Svit - Green World) and the Green Party (Partiya Zelenykh Ukrainy - PZU) on the national question.[3] These attitudes will be correlated to those reflected in a statement issued on nationalism by the Green Parties of Georgia, Lithuania, Ukraine, Estonia and the Armenian Green Movement indicating that the Ukrainian Greens are in line with the view of Greens in other parts of the former Soviet Union in their perception of the national question.

The Greens and the issue of Ukrainian sovereignty/independence

General background

Zelenyj Svit initially stated that it was a nonpolitical green movement whose only concern was with the environment. Rather than divide people along ideological lines, it would seek to unite people for a higher course, namely that of bringing an end to environmental destruction in Ukraine. However, the Greens were soon forced to realise that the environment and politics were inextricably linked. As a result, the Greens had to take an active part in the political process to influence the decision making process directly, in addition to putting pressure on the decision makers through ad hoc activities. It was therefore decided to set up a Green Party, as the "political wing" of Zelenyj Svit.

However, the Green Party in its early days was very cautious. Yuri Shcherbak, its first leader, at one point even claimed that the PZU should not be considered an opposition party in the traditional sense, as "we cannot possibly talk about a confrontation with the Communist Party".[4]

The political implications of environmental reform were brought to the surface by the debate on economic reform. The Stalinist command administrative economic system aimed to maximize economic growth. By stressing quantity and output, hardly any room at all was left for quality or inovation which facilitates the inefficient use of natural resources and, consequently, causes pollution. The centralist character of the Soviet economy further aggravated the environmental costs of industrial pollution.

Most sources of pollution in the republics were controlled by ministries and departments in Moscow. These were not particularly concerned with the environmental aspect of their activities as fines for pollution were set ridiculously low and control of emissions was poor. Their major concern was to fulfil production targets set in their annual production plans. A failure to reach these targets had serious consequences through, for example, a loss of bonuses.

The Chernobyl accident on 26 April 1986, and all the secrecy with which it was surrounded, highlighted the need for more "glasnost" not only on the environment but in Soviet society as a whole. Besides, the authority of the Communist Party had been considerably eroded, due to its inability to take the measures required to minimize the impact of nuclear fall out on the population and on the environment. The accident created a more general awareness in the Soviet population of the link between environmental pollution and health and it also demonstrated the helplessness of individual republics faced with environmental disasters inflicted upon them by Moscow through industrial and energy policies upon which they could exert limited, if any, influence. Furthermore, they were given only restricted access to data (kept in Moscow) revealing the extent of the accident in terms of contamination of the soil as well as in terms of the direct impact on people's health. Besides, it was difficult to uncover the full impact of the accident, given that information was gathered by a large number of bodies and not collated afterwards.

Implications for Ukraine

After the Chernobyl accident ecoglasnost gained momentum, and the impact of pollution generally on the individual republics became better known. In the case of Ukraine, Shcherbak was able to reveal, 25 per cent of the USSR's gross national product was being created in this area and pollution in Ukraine similarly accounted for some 25 per cent of the Soviet Union's total pollution despite the fact that it covered only 3 per cent of the Soviet Union's total territory.[5] In addition, Ukraine was a key republic in the Soviet nuclear power programme, accounting for 25 per cent of its nuclear reactors, which in turn accounted for 40 per cent of the Soviet Union's nuclear capability. In 1986, five nuclear power stations were in operation at Chernobyl, Zaporozh'e, Yuzhno-Ukrainsk, Rovno and Khmelnitsky. Two were being built at Chigirin and on the Kerch peninsula in the Crimea. Two combined nuclear power and heating stations were being constructed in Kharkov and Odessa. In 1986 nuclear energy accounted for 22 per cent of Ukrainian electricity. By the year 2000 this was meant to rise to 60 per cent. Ukrainian scientists in late 1986 at a conference on nuclear power concluded that 90 per cent of Ukrainian territory was unsuitable for nuclear power.[6]

Despite the accident at Chernobyl, the Soviet leadership initiated a series of industrial and energy generating projects in late 1986 and early 1987. In Latvia, plans were made to build a hydro electrical power station on the Daugava river, running through the capital, Riga. In Estonia (Kohtla-Jahrve) an extensive phosphorite mining operation was proposed which many feared would infect the ground water and damage large areas of arable land. In Armenia, Lithuania and Ukraine nuclear power schemes continued unchanged. Construction work at the Armenian nuclear power station and the one at Kerch in the Crimea was started despite the warnings of scientists that they were located in seismically active areas and that an earthquake could cause a serious accident.[7]

The traditional link between nature and nation

In the Baltic states, Armenia and Ukraine in particular, the question of environmental protection eventually became a question of physical survival not only for the people living there, but also of the land on which they lived and of the cultures that had been developed by these people on this land over the centuries. Terms like "genocide" and "national destruction" were used to describe the ultimate effect of Moscow's policies towards the republics.

Ecological reform was linked to economic and political reform more generally, the argument being that a healthy environment was a prerequisite, accompanied by the restoration of national languages as official languages in each republic and the rebirth of national culture and alternative values, for national survival.

In Slav cultures the relationship between the nation (the ethnic group and the land on which they live) and nature is closely linked, not only emotionally as reflected in their literatures and cultural traditions but also etymylogically. It is interesting to note that the words in Russian for Motherland, People (Nation) and Nature all have the same etymological root, namely rod, which means birth, origin: rodina, narod and priroda. The equivalent in the Ukrainian language is rid/rod, which can be translated as lineage, descent, origin. Ridni krai (one's native land, motherland), narid/narod, and priroda in Ukrainian correspond to the Russian words quoted above. In addition, the Ukrainian word for family/kin is rodina. This link finds its expression in the following quote by the Russian writer, Mikhail Prishvin: Liubit "rodinu, znachit berech" prirodu! (To love the Motherland is to protect nature!).[8]

Also, the Orthodox Church emphasises the close relationship between man and nature to a much greater extent than the Protestant and the Catholic churches do. In his standard work on the Orthodox Church Bulgakov (1935) has expressed this relationship as follows:

82

The Holy Spirit is extended by the Church over all Nature. The destiny of Nature is allied to that of Man; corrupted because of Man, she awaits with him her healing. Our Lord, having taken on himself the humanity, has joined his life to all of nature. He walked on this earth, he looked at its flowers and its plants, its birds, its fish, its animals. He ate of its fruits. He was baptized in the water of Jordan, he walked on its waters, he rested in the womb of the earth, and there is nothing in all creation (outside of evil and sin) which remains to his humanity. So the Church blesses all creation...[9]

The cult of the Saints, which occupies a significant place in the Orthodox Church, also has an "ecocultural" side to it. The Saints are considered to be Man's friends, praying with him and protecting not only Man, but also the cattle from wild animals (the Holy Vlas and Modest). Horses are protected by Saint Georgiy, whereas Kuz'ma and Demyan care for the hens and farmers can turn to the Holy Ilya for rain, thus securing a good harvest.[10] In village churches these Saints are often portrayed on icons. Further, a tradition of blessing fields, trees and cattle has survived.

Finally, the close relationship between religion and nature is reflected in a large number of religious feasts celebrated by the Slav Church, directly linked with the cycles of nature. These are observed not only in the countryside, but also by churches in big cities like Kiev. Best known is "Troitsa", celebrated on the 7th Sunday after Easter and marking the beginning of spring.

The link between nation and environment, then, was not a result purely of the political situation in the Soviet Union. It was triggered by it, but it is also deeply rooted in Slav culture. One should also remember that the first informal groups to emerge in the former Soviet Union were concerned with ecocultural issues.

Probably the best known is the movement to save the Hotel Angleterre in Leningrad from being pulled down in early 1987: the famous Russian writer Sergei Esenin committed suicide in this hotel. Esenin represented the so called derevenshchiki, immensely popular in Russia at the beginning of this century. To them nature had intrinsic value and they stressed the close relationship between Nature and Man. Esenin grew up in the Ryazan district and, dressed in peasant clothes, read his poetry in the literary salons of Russian cities with great success. However, he found it very difficult to adapt to the life of "developed society". At the beginning of the 1920s he married the famous dancer Isodora Duncan and went to live with her in the United States. The marriage broke down and

Esenin returned to live in Russia. Disillusioned with the new society emerging in the aftermath of the revolution, with its stress on industrialization and urban development, Esenin finally committed suicide in 1925.

Ecoculture and the Green movement in Ukraine

Although the establishment of a Green movement was very much a response to Chernobyl and the deteriorating state of the environment more generally, some of the first informal groups to emerge in Ukraine were concerned primarily with culture and the environment. It was argued that culture is very closely linked to the environment. A culture can be preserved and developed only within a healthy natural environment. If this environment is damaged, it will inevitably pose a threat not only to areas of cultural significance, but also to the people living in this environment. It is understandable, therefore, that groups to promote Ukrainian culture also took a keen interest in preserving the natural environment.

In the spring of 1987, Tovarystvo Lev (the Lion Society) was set up in Lviv. It described itself as an "independent, community eco-cultural youth-organisation",[11] thus clearly stating its committment to the traditional link between culture and the environment. The Society was organized around four sections: History, Ethnography, Ecology and Social politics.

The Ukrainian Culture and Ecology Club (known as the Culturological Club), which was formed in August 1987 in Kiev, was primarily concerned with those aspects of Ukrainian culture which also had a national aspect.[12] Nuclear power and the environment were thus linked to the survival not only of Ukrainian culture but also to the survival of Ukraine as a nation.[13]

The nationalist movement (RUKH) was also concerned with the link between ecology and national survival. Mykhailo Horyn, the chairman of RUKH's Secretariat, addressing the National Endowment for Democracy in Washington DC in 1990, expressed this concern in the following way: "Ecological consciousness became part of our national consciousness...demonstrations against nuclear power were part of the larger protest against the (Soviet) empire itself".[14] However, whereas independence for the national movement was a goal in itself, for the Greens it was primarily, as we shall see below, a means by which to facilitate an improvement in the state of the environment in Ukraine.

Zelenyj Svit and the Green Party of Ukraine: the national question and the environment

The Greens, like virtually every political movement and party in Ukraine in the late 1980s/early 1990s, called for Ukrainian sovereignty/independence. Their stand on the national question was developed through a series of steps: first, they tried to identify the sources of environmental pollution in the USSR; secondly, they examined the impact of this pollution; and thirdly they offered a solution to the problem: namely economic and political independence.

Section Two of Zelenyj Svit's programme thus addresses the causes of environmental degradation. It states that "the centralist character of the Soviet economic system as well as its stress on economic growth, facilitated a criminal exploitation, a plundering of nature by all-union and republican bureaucratic structures. The ecological crisis has reached catastrophic proportions and threatens not only Nature but also Ukraine with extinction - not only the territory of Ukraine, but Ukraine as a nation".[15]

In this connection, several references are made to the historic and ecocultural legacy of Ukraine. Zelenyj Svit states that its major priority is "the native Ukraine, the Fatherland of the peoples who inhabit this ancient Slavonic land". Nature - the steppes, forests and mountains - are described as "the beauty and the pride of Ukraine".

Shcherbak, when addressing Zelenyj Svit's First Congress on 28 October 1989, spoke about the environmental destruction of Ukraine in very emotional terms:

> Ukraine is facing an ecological catastrophe! Once a flourishing and blessed country with a charming Nature, extolled by generations of our ancestors, Ukraine is now getting close to an ecological apocalypse. We must gather all the healthy forces in our Nation in the fight for SURVIVAL.[16]

He finished his speech with the following words: "Ukraina bula! Ye! Bude! Narod zhytume. Na svoiy zemli. Vichno". (Ukraine was, Ukraine is, Ukraine will be! The people is alive. On its own Land. Forever).[17]

This link between Nation and Nature was to be deepened on later occasions.[18] The thrust of this issue is that when a nation no longer feels that it is the master of its own land the sense of responsibility for this land is lost, and the result is, amongst others, environmental degradation. Only by reimbuing people with a love for their historical cultural heritage and for their native land is it possible to save not only the nation but also the physical territory it occupies.

The Greens thus acknowledged Ukraine's right of existence as a nation. Compared to nationalist forces which demanded outright independence, though, the Greens were more cautious when discussing this issue. In a speech recorded in the Green's newspaper *Zelenyj Svit* in May 1990, Shcherbak drew a restrained conclusion from the view presented above, stating that the only way in which Ukraine could successfully fight for survival was to be given direct responsibility for its natural resources:

> I think that already today it would be sensible to adopt if not a law, then at least a declaration of Ukrainian sovereignty with all its implications, i.e. international- legal, economic, political, ecological, and financial sovereignty. Until we solve the problem of Ukrainian sovereign control over industrial objects, it is impossible to think of any improvement in the ecological situation.[19]

After the Declaration of Sovereignty was endorsed by the Ukrainian parliament in June 1990, Shcherbak discussed the implications of Ukrainian sovereignty in an interview with Ukrainian radio.[20] When asked whether the Greens now favoured self rule, he responded that there could no longer be any doubt on this issue as there already existed the 1990 Ukrainian declaration of sovereignty, expressing the wish of the Ukrainian people. However, at a session of the USSR Supreme Soviet he had suggested the creation of a system of collective security in the military, ecology and energy fields set up horizontally between sovereign states (former Soviet republics).

On numerous occasions representatives of Zelenyj Svit stressed that a prerequisite for improving the state of the environment in Ukraine was the right of control of its own resources. However, the extent and severity of environmental damage on the former Soviet territory was such that it could be addressed successfully only by all the republics together. Initially, at least, a loose federation of former Soviet republics was thus favoured by the leadership of Zelenyj Svit and the Green Party.

As can be seen from the quotes above, officially the Greens have gone to great length to justify their position on Ukrainian sovereignty in terms of care of the environment. This, however, does not mean that Zelenyj Svit and the Green Party are completely devoid of nationalist elements, using the green cause to advance other political goals. It is possible to identify two nationalist currents amongst the Greens: a Ukrainian one and a Russian one. The first manifested itself in the draft programme of the PZU developed by Oleksandr Svirida from Kiev. The programme was discussed at the First

Party Congress, but was eventually rejected. The second caused considerable problems for the Crimean greens following the declaration of Ukrainian Independence of August 1991. The central leadership of both the Green Movement and the Green Party, though, reject the arguments used by these groupings, maintaining that the views of some particular activists are in no way representative of the general mood of the Movement and the PZU.

In the case of Zelenyj Svit, people whose views on the national question differ radically from those held by the Movement, have either been excluded or have left themselves, thus "solving" the issue. Two members of the Kiev branch, for instance, were excluded from Zelenyj Svit in the autumn of 1992 after having written an article which claimed that Ukraine's environmental problems were caused by the Jews and Russians. A group headed by the founder of the Crimean Green Movement, Sergei Shuvainikov, and favouring Crimean reunification with Russia, eventually left the Greens to join the Russian Party.

Although the nationalist grouping withing the Green Party created considerable problems for the party leadership, it chose a different approach, making an effort to achieve a consensus on the national question. This question was debated at length during the First Party Congress in Kiev. A summary of the Congress, given to the author by Kononov, suggests that "a large majority" of the delegates from the Western oblasts of Ukraine (Ternopil, Lviv Ivano-Frankivsk and Zakarpatya) supported the idea of a federal Ukraine, acknowledging the substantial historical and cultural differences between the various parts of Ukraine. The Carpathian delegation even went so far as to state that Carpathia "'never belonged to Ukraine". A conciliation committee headed by a USSR People's Deputy, Leontiy Sandulyak, was given the task of mediating between the "nationalists" and the "internationalists". In the end the following view prevailed:

> If one gives priority to national revival/rebirth rather than ecology, then it is easier to join up with one of the already existing parties and senseless to create a new green one...[21]

Consequently, the Green Party is not hostile to nationalist sentiments, but only in as far as these can be utilized in a positive manner to secure the protection, and in the long run also the improvement, of the natural environment. Although a compomise was reached, the national issue remained an issue of some dispute within the Party. A Conference on National Policies was therefore held shortly afterwards, in Kiev, on 20 January 1991. The Conference adopted a resolution which sythesises the national ethos with the survival of man and indirectly therefore also the survival of the natural environment. The idea of Ukrainian federalism held by the West Ukrainians was also endorsed:

> the PZU holds the opinion that the stable future of Mankind and the renewal of a harmonious coexistence between Man and Nature are possible only provided that the uniqueness and the ecological value of any ethos of a nation, the preservation of all its diversity and the granting to each nation the right of independence are recognised. It is possible to attain all this on the territory of the USSR only provided that authority is decentralised and a complex of democratic reforms implemented in the colonial empire. We think that the first and essential step in this direction is the building of a sovereign Ukraine's real statehood...the PZU favours the renewal of a future federative Ukraine, which will guarantee the harmonious development of all its regions.[22]

Justification: the principle of decentralization

The Ukrainian Greens have tried to justify their position on the national question to the international green community also by referring to traditional green concepts. They claim that their views are in line with the green principles of decentralization and diversity. Moreover, it is argued that rather than violating the principle of "think globally, act locally", this is exactly what they are in practice doing by favouring Ukrainian independence.

For the Greens, the Ukrainian nationalist movement, at least in its early stage, "is absolutely progressive" as it is aimed "not against other nations, but against an empire, the last of its kind in our world..."[23] The Soviet past is rejected as something horrid forced upon the people of the USSR with devastating effects, particularly in terms of the environment:

> more than 70 years of totalitarian rule which was created on the ruins of the Russian empire - "the people's prison" - and continuing its traditions, has put Ukraine on the verge of catastrophe. The ruin of the natural and the cultural environment which reached a peak in the Chernobyl accident facilitated the popular democratic movement in the name of survival of Nature and Man.[24]

88

Therefore the Greens cannot possibly not endorse this movement, although in developed, democratic societies similar coalitions are impossible.[25]

Moreover, the Ukrainian Greens are not against cooperation with Greens in other republics and between republics on environmental issues. PZU thus cooperates closely with the Green Parties of Georgia, Lithuania, Estonia and the Armenian Green Movement and has on a number of occasions issued joint statements with them. Such a statement was issued on the national question in September 1991. In the statement the Greens try to conceptualize their stand on the national question in terms of ecology:

> the Greens' understanding of sovereignty follows naturally from the ethno-ecological principles of diversity, identity, and the eco-system, expressing also the right to life and self-determination of ethnic groups and nations in order to carry out their ecological responsibility, to control their own destiny and to secure a green path of survival and development.

The highest expression of sovereignty is state sovereignty and complete independence of the nation, this nation having its own historical community in the form of a culture, a language and a distinctive relationship with Nature and Life in the region it inhabits.[26]

The declaration concludes that State sovereignty is an inalienable natural right and an expression of true human rights, to which the Greens are deeply committed. It is also the belief of the greens that environmental problems affecting more than one republic can more effectively be addressed between national and state formations than between smaller units within an empire.

National independence and the minority issue

The Ukrainian Greens are deeply committed to the International Declaration of Human Rights and favour a democratic Ukraine, giving extensive rights to all ethnic groups living on its territory.[27] It is made very clear that the majority must not harm any minority's rights. The PZU Manifesto states very clearly that the Greens are opposed to "dividing Ukraine into separated societies, creating discord between those people who live on our land", then , as pointed out by Yuri Shcherbak, "all of us in the West and East, in the South and North" suffer equally from pollution, and "all of us - leftists as well as rightists have identical genetic systems". Therefore the Greens seek to "unite all peoples of Ukraine...in the idea of the survival of Humanity, the nation, and ecological revival in the future".[28]

Harmony is a key concept for the Greens, who eventually hope to create what they call an ecosolidaric community "where the interests of the individual as well as the interests of every ethnic and social group and every nation unite with the highest (ultimate) laws of Nature".[29] To the extent that there is disagreement between political groups in society, consensus should be sought where possible.

National independence and the relationship to neighbouring states

Ukraine has recognized all the new independent states emerging from the former Soviet Union. It has also made it clear that it has got no territorial claims either towards these countries or towards any of its neighbouring countries to the West. Ukraine favours extensive cooperation with the former Soviet republics, also on the environment, to make the transition towards capitalism smoother. However, there is considerable scepticism of the CIS as this is thought to be a continuation of the former Soviet hegemony under Russian rule. Consequently, Ukraine has decided not to sign all agreements made within the framework of the CIS, including an agreement on environmental protection signed in February 1992.

The Greens were against joining the CIS on the grounds that it allowed Russian hegemony over the other former Soviet republics. The statute of the CIS, for instance, was labelled a "diplomatic game of Russian politicians"[30] and was thus unacceptable. Ever since Ukraine joined, the PZU has consequently called for the withdrawal of Ukraine from the CIS framework. The Ukrainian Greens have also expressed their anger towards Russia for claiming with some kind of natural right that which belonged to the Soviet Union and which should be shared by all the former Soviet republics. There is also considerable fear of what has been referred to as "Russian chauvinism".

In a statement issued to the Russian parliament in 1992 the Greens denounced Russian claims to the Crimea as interferance in internal Ukrainian affairs. Also, they rejected the Russian demand that it be given half the Black Sea Fleet with a right to operate it from the Ukrainian port of Sevastopol', as "Ukraine has got the right to possess everything which is located on its territory".[31]

Although the Greens consider themselves to be "anti militant" and have called for a withdrawal of troops from the Western parts of the republic on the grounds that neighbouring countries might perceive these as a threat and have also called for the destruction of all nuclear weapons in

Ukraine and a ban on the production of any kind of mass destructive weapons, they still favour a Ukrainian army, on the grounds that such an army will not pose a threat to anyone, given that Ukraine has not laid claims on any other state's territory.[32]   Such an army is required to protect Ukraine from possible attacks from Russia and to deter Russia from ever considering such an attack. Given the tense relationship that has prevailed between Russia and Ukraine for some time, even Greens have been sceptical about sending Ukrainian nuclear warheads to Russia to be dismantled. Russia is not considered to be reliable in its relationship to Ukraine.

Conclusion

In summary, then, in so far as one chooses to refer to "Green nationalism" this is a tolerant nationalism in that it allows for ethnic minorities and the granting of extensive rights to these minorities. It is a constructive nationalism in that it wants to rebuild what a "dictatorial empire" has destroyed. It is anti Russian only in as much as there is fear amongst Greens that Ukraine will develop into a Russian satelite.[33]

The Greens' stand on Ukrainian independence was motivated solely out of concern for the environment.   Any nationalism that cannot be justified on Green grounds is rejected by Zelenyj Svit as well as the PZU (against West-Ukrainian independence, Crimean reunification with Russia). Territorial expansionism is denounced outright.   On the occasion of the war in Yugoslavia, PZU has issued a statement against Serb and Croat expansionism in Bosnia.

"Green nationalism", deeply rooted in traditional Slav ecoculture, is as far as we can see perfectly compatible with a more global commitment. The fact that the PZU is a member of the European Greens and that they are working closely with Green Parties and groups in Western Europe gives further evidence of this commitment.

When referring to the former Soviet Greens as nationalists, we should remember that the Western Greens emerged in stable democratic national states which had existed for a long time and whose right of existence was never questioned. Greens in the West should acknowledge these differences. Also, they should realise that because the political context within which the former Soviet Greens operate is so different, views held by these "Soviet Greens" may differ from their own, but be motivated by a commitment to the very same international Green principles.

Notes

1. Goldman, M.(1992), "Environmentalism and Nationalism: an unlikely twist in an unlikely direction" in Stewart, J.M. (ed.), *The Soviet Environment: Problems, Policies, and Politics*, Cambridge University Press, Cambridge, pp. 1-10.

2. Interviu pered viborami (28 February 1992), "Zeleni" problemi - dlya sitoho suspil'stva, *Molod Ukrainy*.

3. For background information on the Ukrainian Greens, see for example Marples, D.R. (1991), *Ukraine under Perestroika- Ecology, Economics and the Workers' Revolt*, Macmillan, London, Chapter 5 or Khronika Zelenoho Svitu (March 1991), *Zelenyi Svit*, no. 4.

4. Zeleni - shche zeleni (16 February, 1990), *Radyans'ka Ukraina*.

5. Interview with Shcherbak (August 1991), Kiev.

6. (April 1991), *Ukrainian Reporter*, vol. 1, no. 8, p.1.

7. ibid.

8. quoted in Ozhegov et al.(1991), *Ekologicheski impul's*, Iz-vo Nauka, Moscow, p.146.

9. Bulgakov, S. (1935), *The Orthodox Church*, The Centenary Press, London, pp. 157-58.

10. Nozova, G.A. (1975), *Yazychestvo v pravoslavii*, iz-vo Nauka, Moscow, p.78.

11. Kuzio, T. (1990), *Restructuring from below. Informal Groups in Ukraine under Gorbachev, 1985-90*, Ukrainian Press Agency, London, p. 6.

12. ibid. p.4.

13. ibid. p.5

14. cited in Feshbach & Friendly Jr. (1992), *Ecocide in the USSR: Health and Nature under Siege*, Aurum Press, London, pp.232-33.

15. (October, 1989), *Zelenyj Svit*, Prohama, section 2, Kiev.

16. Shcherbak, Yuri (April, 1990), "Bozhe vryatui Ukrainu", *Zelenyj Svit*, no.1, p.2.

17. Shcherbak, Zhiti (28 October, 1989), *Zelenyj Svit*, p.2.

18. cf. Oleksandr Svirida in interview with Ukrainian Radio (K-3, 22:30, 19 January 1991) cited in Radio Liberty's *Ukraine Today* or see Suverenitet, "Ekologia i politika zelenykh za vyzhuvannya", *ZVU*, 12 March, 1991.

19. Urochisto proholoshuemo stvorennya Partii Zelenykh Ukrainy -Manifest PZU, presented by Shcherbak and (May, 1990), *Zelenyj Svit*, no.2, p.2.

20. Interview with Ukrainian Radio (K-3, 1:30, 13 December 1990), quoted in Radio Liberty's *Ukraine Today*.

21. summary of First Congress of the PZU provided by Vitali Kononov.

22. (20 January, 1991), *Resoljutsia konferentsii PZU z natsional'noy politiki*, Kiev.

23. summary of First Congress of the PZU provided by Vitali Kononov.

24. Prohramny printsipy PZU, endorsed on 15 July 1990.

25. summary of First Congress of the PZU provided by Vitali Kononov.

26. (September, 1991), "Sovmestnoye zayavlenie Partii Zelenykh Gruzii, Litvy, Ukrainy", *Estonii i Dvizheniya Zelenykh Armenii*, Tbilisi.

27. (12 March, 1991), "Suverenitet: Ekologia i politika", *ZVU*, p.1.

28. see (May, 1990), "Manifest PZU", *Zelenyj Svit*, p.2 and (September, 1990), "Virimo v zelenu revoljutsiju", *Zelenyj Svit*, no.11, p.2.

29. (October, 1989), First Programme of *Zelenyj Svit*, endorsed by the First Congress.

30. Zeleni vvazhajut', shcho pidpisuvati statut SND Ukraina ne zatsikavlena, (UT-1, 21 January 1993, UUTN, 16:00, 21:00) quoted in Radio Liberty's *Ukraine Today*. Fear that Ukraine would be turned into a Russian satellite is

found in Zayava PZU: hibinna lohika dagomis'khikh domovlenostei vede do vtyahuvannya Ukrainy u vis'koho-politichnu orbitu Rossii, (Radio Ukraina, 21:50, 7 July 1992), cited in Radio Liberty's *Ukraine Today*.

31. Otkritoye pis'mo PZU k Verkhovnomu Sovetu Rossiskoi Federatsii.

32. Zayava PZU v sv'yazhku z nastinimi sprobami rosiskoi derzhavi utverditi svoye osoblive pravo na armiju kolishnoho SRSR ta rozv'yazanu zasobami masovoi informatsii antiukrains'ku kampaniu. See also Zayavlenie PZU kasajushchiesya oboronnikh voprosov.

33. Zayava PZU: hibinna lohika dagomis'khikh domovlenostei vede do tryahuvannya Ukrainy u viys'koho/politichny orbitu Rossii (Radio Ukraina, 21:50, 7 July 1992), cited in Radio Liberty's *Ukraine Today*.

3 Conflict over South East Asia's forests: A political ecology perspective

Raymond Bryant

Environmental degradation associated with natural resource exploitation is an issue of growing political controversy in South East Asia. If the dynamic of exploitation and degradation varies from country to country, there is nevertheless an overall pattern which is discernible: South East Asian states have exploited the natural resource base within their jurisdiction in order to promote political and economic objectives. In a sense, this process merely reiterates a pattern of development that was first elaborated on a grand scale under colonial rule. However, the post colonial quest to "modernize", in so far as this has been associated with industrialization, has intensified the pressure on the natural resource base. Whether capitalist or socialist, "outward looking" or "inward looking", democratic or authoritarian, states in South East Asia have shared a similar vision of development predicated on industrial development and intensive natural resource exploitation (Bryant et al., 1993b).

The following paper provides an overview of this process. Specifically, and with reference to Burma (Myanmar), Thailand and Indonesia, it examines the ways in which political and economic forces have conditioned forest conflict and change from the colonial era to the present. The three countries referred to in this paper provide an interesting contrast in terms of the impact of colonial rule, development strategies, and so on; they are useful as case studies precisely because they highlight the general point that, notwithstanding differences, similarities in terms of natural resource exploitation and environmental degradation abound. Given this selective country focus, the paper does not attempt, therefore, a region wide survey (Hurst, 1990; Poffenberger, 1990; Bryant et al., 1993a); nor does it aim to be comprehensive in its coverage of the issues in the three countries under review.

In seeking to understand the connection between environmental degradation and forest exploitation in Burma, Thailand and Indonesia, the approach adopted in this paper is that of political ecology. An emerging research agenda in Third World environmental and development studies, political ecology highlights those political and economic processes which have contributed to environmental change. The goal of this paper, then, is to suggest broad patterns and possible explanations for South East Asian forest change in keeping with a political ecology perspective.

Political ecology

As it pertains to the Third World, political ecology is a research agenda that emerged during the 1980s (Bryant, 1992; Peet and Watts, 1993). Blaikie and Brookfield (1987: 17) offer the following definition:

> The phrase "political ecology" combines the concerns of ecology and a broadly defined political economy. Together this encompasses the constantly shifting dialectic between society and land-based resources, and also within classes and groups within society itself.

Such a definition has much to commend it, particularly when extended to encompass not only "land", but more generally "environment". Political ecology, then, constitutes a research agenda concerned with the political and economic sources, conditions and ramifications of environmental change.

As such, political ecology needs to be differentiated from much of the literature associated with the concept of sustainable development. Following publication of the World Commission on Environment and Development's report *Our Common Future* (1987), a literature developed centred on this important concept that has sought to elucidate its meaning as well as its policy application (Elliot, 1994). As scholars note, however, sustainable development is a rather slippery, chameleon like concept: it means many things to many people and changes colour according to use (Lele, 1991; Redclift, 1991; Murdoch, 1993). Furthermore, the sustainable development literature is premised largely on economics, but it is this discipline which is "at the heart of the problem of why development has been unsustainable" in the first place (Norgaard, 1994: 18). Given the economistic bent of the literature, it comes as little surprise that much of this writing on sustainable development is devoid of consideration of politics. Policies are formulated and implemented seemingly without conflict, and assumptions are made about the role of the state and societal actors that obviate the need for political analysis (Bryant, 1991).

My objective here is not to provide a detailed critique of the sustainable development literature (cf. Redclift, 1987; Adams, 1990; Sachs, 1993; Norgaard, 1994). Rather, it is simply to note that, unlike that literature, research in political ecology is concerned with understanding the role that political and economic forces play in shaping environmental change. The essential point to remember is that from a political ecology perspective, the environment is seen as being in a politicized condition. It logically follows, then, that environmental change is ripe with political meaning.

Colonial ideologies of forest control in South East Asia

The utility of a political ecology perspective becomes more readily apparent when such a perspective is utilized to assess forest conflict and change in South East Asia. To appreciate such conflict in contemporary Burma, Thailand and Indonesia, however, it is first necessary to review the colonial origins of ideologies of forest control.

To speak of ideologies of forest control in South East Asia is effectively to describe a long, and conflict ridden process whereby the state asserted centralized control over the forests. In certain cases, this process predated colonialism. In monarchical Burma, for example, commercially valuable teak forests in central and southern regions of the country were regulated from at least the mid eighteenth century, if not before (Bryant, 1993). In other cases, as in Java, the assertion of state forest control was more closely linked to the vicissitudes of a lengthier period of colonial rule (Peluso, 1991; Boomgaard, 1992a).

If the origins of state forest control in South East Asia are thus variable, the late 19th and early 20th centuries witnessed a general extension and "deepening" of such control. Scientific and administrative techniques were pioneered and elaborated. As "Leviathan" grew in size and became more institutionally complex, it was able to extend its sway over a progressively larger area of forest lands. Just as "scientific" forestry was imported from Germany and France to facilitate the commercial development of selected species, so too functionally defined forest departments were established (for example, Netherlands Indies in 1849, British Burma in 1856, Siam in 1896) to enforce this powerful new ideology of control (Boomgaard, 1992b; Bryant, 1993; Mekvichai, 1988).

The aims of state forest control centred on diverse political and economic calculations. In Burma, the British were primarily concerned with the extraction of teak for military strategic and commercial reasons (Bryant, 1993). Beginning in the early 19th century, teak was an essential timber in the construction of military and civilian vessels; it was also for a time a source of railway sleepers. Thus, both at sea and on land teak was an invaluable resource in the "stitching together" of empire and its ensuing commercial prosperity. Similarly, in Dutch ruled Java, teak was important for imperial economic development. During the so called Cultivation System of the mid 19th century, teak was extensively harvested for use in the construction of sugar factories, coffee warehouses and plantation housing, as well as for fuel in the processing of sugarcane, coffee and tobacco (Peluso, 1991). From the late 19th century Siam's rulers asserted control over the teak forests of northern Siam partly as a means of deriving income from that lucrative trade.

But as the Siamese example also illustrates, the assertion of state forest control could also be motivated by strictly political considerations (Mekvichai, 1988). Beyond the question of timber revenue, Siam's leaders sought to assert central control over the peripheral areas of the kingdom at a time when the British and French were extending their grip over the region. The British annexation of Burma, between 1824 and 1886, illustrated the tenuous nature of Siamese independence and prompted the Siamese to embark on a "modernisation" programme which included territorial delimitation and control of the northern teak forests (Vandergeest and Peluso, 1993).

There were strictly political, as distinct from commercial, advantages to be gained by the Dutch and British from the assertion of state forest control. In Java and Burma, as elsewhere, the forests were traditionally the home of those fleeing or challenging central authority (Adas, 1981). This tradition persisted under colonial rule, and one, albeit indirect, function of forest management was to deny these forested realms to the state's opponents (Peluso, 1992a; Bryant, 1993).

Whatever the motivation, the assertion of state forest control was met by the fierce resistance of those whose access to forest resources was thereby limited or denied. In Java, rules brought in by the Dutch declaring all forest land and teak to be government property led to a confrontation with villagers who persisted with wood gathering and grazing practices that were now illegal (Boomgaard, 1992b). That confrontation was particularly severe in teak forest villages where forest police considered even "the smell of teakwood...as evidence of punishable theft" (Peluso, 1991: 72). In Burma, the British encountered similar difficulties in protecting reserved forests and species, particularly in low lying forests that adjoined human settlements (Bryant, 1993). In Siam, where state forest control was less rigorous, such conflict was less typical. However, in the early 20th century access restrictions increased, as did local peasant resistance (Mekvichai, 1988).

If, as a result of new or more strictly enforced regulations, the colonial era was marked by the intensification of conflict over forest access it was also characterized by the deforestation of vast low lying areas suitable for wet rice agriculture. Thus, at the same time as states formally limited popular access to commercially valuable forests, they encouraged peasants to convert other forests to permanent agriculture (Rush, 1991). During the late 19th and early 20th centuries, much of central Siam, Lower Burma and Java were cleared of forest in this manner. In Burma, for example, more than three million hectares of mangrove swamp and *kanazo* forest were eliminated to make way for paddy cultivation (Adas, 1983). Lower Burma became the leading rice exporting region of the world in the late colonial era.

All of these developments resulted in a dramatic transformation of the forested landscape during colonial times. Notwithstanding sharp peasant resistance, the state asserted ever greater control over forest lands. In some areas, such control was manifested in terms of the retention of forest cover; in other areas, it was marked by widespread forest clearance in aid of increased agricultural production. This combination of state control, popular resistance and deforestation has persisted in the post colonial era.

Post colonial forest conflict and change

Following the Second World War, Burma and Indonesia gained independence. Concurrently, Siam (or Thailand as it was by then called) was able to shake off residual elements of informal colonial control. The three states assumed formal political control over their territories, and thereby over the fate of the forests within those territories.

During the next four decades, the total area under forest in these countries declined (Table 1).

Table 1

Forested Area: Burma, Thailand and Indonesia
(Area in mn ha; % = % of total land area)

	Burma		Thailand		Indonesia	
	Area	%	Area	%	Area	%
late 1940s	>50.0	75	32.6	69	-	-
early 1960s	38.6	59	27.1	53	-	-
early 1980s	27.6	42	13.0	25	>157.0	80
late 1980s	25.7	39	7.7	15	117.9	62

Source: Blower et al., 1991; Hirsch, 1987; Gillis, 1988; Cox et al., 1991.

What is of interest here is less the absolute numbers than the overall trend as forest related estimates are notorious for their wide variations depending on the definitions used. From a high of 75 per cent after the Second World War, Burma's total forest cover has fallen to under 40 per cent. With about one half of South East Asia's total forest cover, Indonesia's deforestation has not occurred as extensively as in Burma. Yet, if the estimated annual deforestation rate of between 700,000 and 1.2 million hectares per annum in recent decades is accurate, then Indonesia has by far the highest regional level (Gillis, 1988; Cox et al., 1991). In contrast, rapid deforestation in the context of a much more limited Thai forest resource base has led to the elimination of most of that country's remaining tracts. Although officially Thailand's forests cover as much as 29 per cent, in reality when degradation and conversion to agriculture is taken into account, the real figure is 15 per cent (Hirsch, 1987) or lower (Rigg and Stott, 1992). A major step in Thailand's forest transition came in January 1989 when a nationwide logging ban was imposed. However, this ban led Thai forest companies to shift their operations into neighbouring Burma, Cambodia and Laos in order to maintain timber supplies (Rigg and Stott, 1992).

The general picture, then, is one of ubiquitous deforestation over more than forty years as Burma, Thailand and Indonesia have felled forests as part of national development strategies. This situation is not atypical of the region (Beresford and Fraser, 1992; Broad and Cavanagh, 1993; Bryant et al., 1993b), or, for that matter, of most of Europe and North America during an earlier epoch (Mather, 1990). But if the Burmese, Thai and Indonesian experiences conform to a global pattern, what are the political and economic processes at work that have shaped their histories?

First and foremost, states in these countries have built upon colonial and precolonial management techniques and structures in order to extend and deepen their control over forested and peripheral areas. If the "territorialization of national space" (Vandergeest and Peluso, 1993) remains incomplete, it has nonetheless proceeded apace during the past forty years.

This process is most advanced in Thailand where today even remote areas feel the impact of state policies. Most evidently, the country has been tied together through the development of infrastructure. The construction of railways, for example, facilitated the assertion of state authority in peripheral forested areas during the colonial era (Lohmann, 1993). More recently, the development of a road and telecommunications network in the 1960s and 1970s assisted the military in its anti communist campaign, but also served to generally strengthen central control. In practice, the nature of such control is quite complex (Vandergeest and Peluso, 1993). As Hirsch (1990) shows for north west Thailand, however, there has been a proliferation of government departments and agencies in recent years which reinforces state control over people and resources.

The record of territorial definition and state empowerment has been more ambiguous in Burma. On the one hand, Burma's leaders have been acutely sensitive to questions of internal political control. Since the 1962 coup d'etat which brought Ne Win to power, considerable energy has been expended on the elaboration of centralized administrative structures (Taylor, 1987). Following the nationalization campaign of the 1960s, the Burmese state controlled most economic activity in the country, including the forestry sector. Recent economic reforms notwithstanding, Burma's leaders continue to maintain a firm administrative grip on society and resources in central parts of the country. On the other hand, however, the ability of the Burmese state to control forested areas at the periphery has been weak. The mountainous periphery of Burma has been home to a succession of ethnic and ideological opponents of the post colonial Burmese state (Smith, 1991; Falla, 1991). The Burmese army has lately contained these groups in border areas, yet control outside of central areas remains tenuous. Indeed, deforestation in border areas has been just as often an indication of a lack of state authority as it has been of its presence.

The Indonesian state has experienced greater success than its Burmese counterpart in asserting authority over forested and peripheral areas. As with Thailand, this authority has played a central role in the deforestation process. Despite the vicissitudes of Indonesian politics under Sukarno and Suharto, the Indonesian government has elaborated management structures first introduced under the Dutch. In Java, for example, the forest service continues to control the teak forests and punish forest offenders (Peluso, 1992a). In the Outer Islands, and notably in timber producing Kalimantan (Borneo), centralized authority is more recent, but if anything has had a more visible and damaging effect on the forests and its peoples (Potter, 1991; Peluso, 1992b). Even more so than in Thailand, the conjuncture of public/private and civil/military interests has been a factor in extensive and nonsustainable logging practices. Indeed, deforestation has been an integral feature of the territorial definition of the Indonesian polity.

If South East Asian deforestation is associated with state empowerment, it is also linked to other factors such as integration in the national and global economy. In Thailand, upland forested areas have been cleared to make way for cash cropping: between 1971 and 1988 the area under crops such as maize, cassava, sugar cane and kenaf trebled (Rigg, 1993). The rapid extension of cash cropping is the leading contemporary source of deforestation (Rigg, 1993), but legal, until 1989, and illegal logging, plantations and miscellaneous development projects such as dams, golf courses and tourist resorts have exacerbated this process (Lohmann, 1993).

National and global economic integration has also contributed to deforestation in Indonesia. The development of a large scale export oriented logging industry in Kalimantan is a case in point. Hardwood log exports increased rapidly after 1967 such that in the 1970s Indonesia, along with Malaysia, was the world's leading exporter of tropical hardwoods (Gillis, 1988; Ooi Jin Bee, 1990). Beginning in the late 1970s, Indonesia moved into timber processing; more than 100 plywood mills were established and a log export ban imposed in 1985 as part of a drive to increase value added in the forestry sector (Gillis, 1988; Potter, 1991). This industrialization strategy has increased pressure on forest stands as extraction now must meet the "insatiable" demand of local mills (Potter, 1991; Barbier, 1993).

The Burmese situation provides a contrast to the Thai and Indonesian cases. Between 1948 and 1988, Burma was a socialist country. The "Burmese Way to Socialism" was official dogma, and after 1962 led to the virtual delinkage of Burma from the global economy. Instead of export oriented development, Burma followed an involuted economic strategy which changed little over the years, and which observers condemned as the "Burmese way to poverty". Ironically, economic ineptitude may have slowed the country's rate of deforestation. However, as the experience of socialist Vietnam corroborates (Beresford and Fraser, 1992), international isolation does not necessarily lead to sustainable forestry policies. In Burma's case, and notwithstanding the colonial legacy of a comprehensive system of forest management (Bryant, 1993), Burma's forests continue to be depleted as a result of illicit felling and encroachment, shifting cultivation, and more recently short term logging concessions along the Thai Burmese border (Union of Myanmar, 1989, 1990; Harbinson, 1992; Bryant, 1994a). Moreover, as Burma restores links with the global economy, the likelihood that the country's forests will be subjected to the same processes as in Indonesia and Thailand is great.

State empowerment and national/global economic integration, then, have contributed to deforestation in Thailand, Indonesia and to a lesser extent Burma. These political and economic forces have also been the source of intensified social conflict as diverse groups contest access to forests and forest lands. Such conflict is most evident in Thailand where, since the 1970s, a myriad of local level citizen's groups have formed to challenge logging and other development projects (Hirsch and Lohmann, 1989; Leungaramsri and Rajesh, 1992). Although their interests are diverse, a common feature of these groups is the quest for local control over decisions that affect local social and environmental well being. More than merely a question of "NIMBY" (not in my backyard) politics, such conflict is about local empowerment and the adaptation of the development process to fit local needs. It is a livelihood struggle (Redclift, 1987) in which the question of land rights looms large (Lohmann, 1993).

A similar dynamic of control and resistance is to be found in Indonesia. More than 21 million people today inhabit Java's state forest lands, for example, and restrictions imposed on popular forest access by the State Forestry Corporation have been an important source of poverty and social strife in these areas. State efforts to control forest lands, species, labour and ideology have been countered by popular resistance which is land based (squatting), species based (arson), labour based (strikes and slowdowns), and which is grounded in a "culture of resistance" (Peluso, 1992a). Similar cultures of resistance, often with an ethnic basis, are associated with forest related conflict in Indonesia's Outer Islands, notably Kalimantan and Irian Jaya (Hurst, 1990; Peluso, 1992b).

The dynamic of control and resistance in Burma may be divided into two elements. First, and in so far as fighting in the border regions reflects the struggle of local ethnic groups for autonomy, if not independence, from the centre, conflict over the forests in these areas is also a battle for territorial authority, and the rights to timber and mineral revenue. The forty six year long Karen struggle to create an independent state along the Thai border needs to be seen in this light (Falla, 1991; Harbinson, 1992). In contrast, in central areas under firm government control, it is likely that patterns of control and resistance are similar to those that existed during the colonial era. The allegedly destructive practices of shifting cultivators, to take one example, are as much a source of official concern today as they were in colonial times (Union of Myanmar, 1989, 1990; Blower et al., 1991; Bryant, 1994a and 1994b). As Burma integrates with the global economy and pressure on the forest resource intensifies, such conflict will intensify.

In Burma, Thailand and Indonesia, therefore, forest conflict and change has shown much continuity with the colonial past. State efforts to enforce central political control and to promote economic development have, with the partial exception of Burma in the 1960s, resulted in post colonial

103

patterns of forest destruction and conflict strongly reminiscent of the colonial era. During both the colonial and post colonial era, the forests have been a central focus of political attention.

Conclusion

This paper has briefly explored forest conflict and change in South East Asia with reference to Indonesia, Thailand and Burma. It suggested that such conflict and change need to be viewed in light of structures of political control and economic activity largely introduced during the colonial era and subsequently elaborated by post colonial governments. In Indonesia and Thailand, and increasingly today in Burma as well, state empowerment and national and international economic integration have combined in such a manner as to remove control over forest resources from the local level. The response has been the elaboration of resistance strategies by peasants, ethnic minorities and other marginalized groups that are designed to restore decision making to the local level.

The patterns of control and resistance described in this paper are not unique to the countries, or even region, in question (Ecologist, 1993). Nor are these patterns confined to situations characterized by forest loss. Indeed, the rapid growth of eucalyptus plantations to service a global pulp paper industry indicate that conflict will increasingly centre on reforestation, and not deforestation issues (Sargent and Bass, 1992; Grainger, 1993). In Thailand, for example, the "politics of eucalyptus" has already become an issue of national importance (Lohmann, 1991; Puntasen et al., 1992).

In essence, such conflict is about power: who is to control forest land use decision making, and generally, the development process itself. It is the task of those who consider themselves political ecologists to investigate and understand conflict in this way. Political ecologists thus acknowledge that environment and development, wealth and poverty, control and resistance, are inextricably linked; and that to speak meaningfully of "sustainable development" is to first address the need for radical social and political change. Let us, therefore, ponder the question "whose common future" before we even begin to consider the social and environmental implications of the phrase "our common future".

References

Adams, W.M. (1990), *Green development: environment and sustainability in the Third World,* Routledge, London.

Adas, M. (1981), "From avoidance to confrontation: peasant protest in precolonial and colonial Southeast Asia",

Comparative Studies in Society and History 23, pp.217-47.

-------- (1983),"Colonization, commercial agriculture, and the destruction of the deltaic rainforests of British Burma in the late nineteenth century". In Tucker, R.P. and Richards, J.F., (eds), *Global deforestation and the nineteenth-century world economy*. Duke University, Durham, pp.95-110.

Barbier, E.B. (1993), "Economic aspects of tropical deforestation in South-East Asia". In Bryant, Rigg and Stott, (eds).

Beresford, M. and Fraser, L. (1992), "Political economy of the environment in Vietnam", *Journal of Contemporary Asia* 22, pp.3-19.

Blaikie, P. and Brookfield, H. (1987), "Defining and debating the problem", in Blaikie, P. and Brookfield, H., (eds), *Land degradation and society*, Methuen, London, pp.1-26.

Blower, J. et al. (1991), "Burma (Myanmar)", in Collins, N.M., Sayer, J.A., and Whitmore, T.C., (eds), *The conservation atlas of tropical forests: Asia and the Pacific*, Macmillan, London, pp.103-110.

Boomgaard, P. (1992a), "Forest management and exploitation in colonial Java, 1677-1897", *Forest and Conservation History* 36, pp.4-14.

--------, (1992b), "Colonial forest policy in Java in transition 1865-1916", in Cribb, R, (ed.), *State without citizens*, KITLV, Leiden.

Broad, R. and Cavanagh, J. (1993), *Plundering paradise: the struggle for the environment in the Philippines*, University of California Press, Berkeley.

Bryant, R.L. (1991), "Putting politics first: the political ecology of sustainable development", *Global Ecology and Biogeography Letters* 1, pp.164-66.

-------- (1992), "Political ecology: an emerging research agenda in Third-World studies", *Political Geography* 11, pp.12-36.

-------- (1993), *Contesting the resource: the politics of forest management in colonial Burma*, Ph.D. thesis, University of London.

-------- (1994a), "The rise and fall of taungya forestry: social forestry in defence of the Empire", *Ecologist* 24, pp.21-26.

-------- (1994b), "Shifting the cultivator: the politics of teak regeneration in colonial Burma", *Modern Asian Studies* 28, pp.225-250.

Bryant, R.L., Rigg, J. and Stott, P., (eds) (1993a), *The political ecology of Southeast Asian forests: trans-disciplinary discourses*, Global Ecology and Biogeography Letters Special Issue 3 (4-6), Basil Blackwell, Oxford.

-------- (1993b), "Forest transformations and political ecology in Southeast   Asia", in Bryant et al. (1993a).

Cox, R. et al. (1991),"Indonesia", in Collins, N.M., Sayer, J. and Whitmore, T.C., (eds), *The conservation atlas of tropical forests: Asia and the Pacific*, Macmillan, London, pp.141-65.

Ecologist (1993), *Whose common future? Reclaiming the commons*, Earthscan, London.

Elliot, J.A. (1994), *An introduction to sustainable development: the developing world*, Routledge, London.

Falla, J. (1991), *True Love and Bartholomew: rebels on the Burmese border*, Cambridge University Press, Cambridge.

Gillis, M. (1988), "Indonesia: public policies, resource management, and the tropical forest", in Repetto, R. and Gillis, M. (eds), *Public policies and the misuse of forest resources*, Cambridge University Press, Cambridge, pp.43-113.

Grainger, A. (1993), *Controlling tropical deforestation*, Earthscan, London.

Harbinson, R. (1992), "Burma's forests fall victim to war", *Ecologist* 22 (2), pp.72-73.

Hirsch, P. (1987), "Deforestation and development in Thailand", *Singapore Journal of Tropical Geography* 8, pp.129-38.

-------- (1990), *Development dilemmas in rural Thailand*, Oxford University Press, Singapore.

Hirsch, P. and Lohmann, L. (1989), "Contemporary politics of environment in Thailand", *Asian Survey* 29, pp.439-51.

Hurst, P. (1990), *Rainforest politics: ecological destruction in South-East Asia*, Zed Books, London.

Lele, S.M. (1991), "Sustainable development: a critical review", *World Development* 19, pp.607-21.

Leungaramsri, P. and Rajesh, N. (1992), *The future of people and forests in Thailand after the logging ban*, Project for Ecological Recovery, Bangkok.

Lohmann, L. (1991), "Peasants, plantations and pulp: the politics of eucalyptus in Thailand", *Bulletin of Concerned Asian Scholars* 23 (4), pp.3-17.

-------- (1993), "Land, power and forest colonization in Thailand", in Bryant, Rigg and Stott, (eds).

Mather, A.S. (1990), *Global forest resources*, Belhaven Press, London.

Mekvichai, B. (1988), *The teak industry in north Thailand: the role of a natural-resource-based export economy in regional development*. Ph.D. thesis, Cornell University.

Murdoch, J. (1993), "Sustainable rural development: towards a research agenda", *Geoforum* 24, pp.225-41.

Norgaard, R.B. (1994), *Development betrayed: the end of progress and a coevolutionary revisioning of the future*, Routledge, London.

Ooi Jin Bee (1990), "The tropical rain forest: patterns of exploitation and trade", *Singapore Journal of Tropical Geography* 11, pp.117-42.

Peet, R. and Watts, M. (1993), "Introduction: development theory and environment in an age of market triumphalism", *Economic Geography* 69, pp.227-53.

Peluso, N.L. (1991), "The history of state forest management in colonial Java", *Forest and Conservation History* 35, pp.65-75.

-------- (1992a), *Rich forests, poor people: resource control and resistance in Java*, University of California Press, Berkeley.

-------- (1992b), "The political ecology of extraction and extractive reserves in East Kalimantan, Indonesia", *Development and Change* 23, pp.49-74.

Poffenberger, M. (ed.) (1990), *Keepers of the forest: land management alternatives in Southeast Asia*, Kumarian Press, West Hartford, Conn.

Potter, L. (1991), "Environmental and social aspects of timber exploitation in Kalimantan, 1967-1989", in Hardjono, J. (ed.), *Indonesia: resources, ecology, and environment*, Oxford University Press, Singapore, pp.177-211.

Puntasen, A., Siriprachai, S. and Punyasavatsut, C. (1992), "Political economy of eucalyptus: business, bureaucracy and the Thai government", *Journal of Contemporary Asia* 22, pp.187-206.

Redclift, M. (1987), *Sustainable development: exploring the contradictions*, Methuen, London.

-------- (1991), "The multiple dimensions of sustainable development", *Geography* 76, pp.36-42.

Rigg, J. (1993), "Forests and farmers, land and livelihoods: changing realities in Thailand", in Bryant, Rigg and Stott, (eds).

Rigg, J. and Stott, P. (1992), "The rise of the Naga: the changing geography of Southeast Asia, 1965-1990", in Chapman, G.P. and Baker, K.M., (eds), *The changing geography of Asia*, Routledge, London, pp.60-99.

Rush, J. (1991), *The last tree: reclaiming the environment in tropical Asia*, Asia Society, New York.

Sachs, W. (ed.) (1993), *Global ecology: a new arena of political conflict*, Zed Books, London.

Sargent, C. and Bass, S. (eds) (1992), *Plantation politics: forest plantations in development*, Earthscan, London.

Smith, M. (1991), *Burma: insurgency and the politics of ethnicity*, Zed Books, London.

Taylor, R.H. (1987), *The state in Burma*, C. Hurst, London.

Union of Myanmar, Ministry of Agriculture and Forests (1989), *Forestry situation in Myanmar*, Ministry of Agriculture and Forests, Yangon.

-------- (1990), *Watershed management in Myanmar*, Ministry of Agriculture and Forests, Yangon.

Vandergeest, P. and Peluso, N.L. (1993), *Fixed property in national space: territorialization of the state in Siam/Thailand*, unpublished manuscript.

World Commission on Environment and Development (1987), *Our Common Future*, Oxford University Press, Oxford.

Part III
PLANNING FOR SUSTAINABILITY

1 An examination of current issues in environmental planning and policy in Japan

Ali Parsa and William Penrice

Introduction

> A quality of this people which has been praised by
> every visitor to Japan is their love of beauty. This,
> indeed, runs through the whole of their life,
> wherever it remains untouched by industrialism.
> (Allen G.C. 1927).

Japan's post war development has been through major
transformations as a result of ferocious economic growth and
a rapid population growth. The treatment of environmental
issues has placed Japan under scrutiny both at national and
international levels (Kazu 1993). The treatment of the
environment is an interesting issue for those concerned with
Japanese society in general and environmental issues in
particular. As the only non western society to industrialize
fully, Japan has in the last decades become a vital case study
of industrialization. However, most of these studies have
tended to concentrate on Japanese business practices and
technology development. It is therefore significant to view
how Japan has adapted to the problems of industrialization.
The "Japan Model" or the "economic miracle" became the
model for developing countries to emulate (Mouer &
Sugimoto, 1990). As such, Japan has an important role in
the prevention of environmental degradation in nations
attempting to follow the Japanese example. This is
particularly the case in Asia Pacific countries.

The aim of this chapter is to examine the Japanese
approach to environmental planning and management and
sets out to determine whether there is any room for
environmental considerations in decision making at local and
national levels. Furthermore the chapter will provide an
analysis of the growing environmental debate in the Japanese
context. It will aim to establish the importance of grass root

environmental pressure groups and their impact on the local planning decision making process in the face of pressure from the large construction and real estate corporations. It draws on the results of research and indepth interviews with a number decision makers in charge of large conglomerates, as well as central and local government officials.

Japan's transformation into a post industrial society reveals a fascinating policy making structure as the political system has attempted to incorporate environmental discontent since the 1960s. The importance of new values, changing technology and a demassified political system hold vital clues in the implications for the future of the environment and democracy in Japan (Edmunds, 1983). It is particularly relevant to look at this in the 1990s with a global environmental crisis and a Japan searching for a new international role having a cross cultural validity in the environmental planning and policy field.

A demographic assessment of Japan

Any assessment of development in Japan must observe Japan's geographic and demographic situation. It is therefore important to examine the basic indicators including population, land use and GNP in relation to land use. Japan's limited space, a mere 0.3 per cent of the world's land area, is further constrained by the large percentage of land unfit for development, around 70 per cent. A brief international comparison reveals some interesting statistics and helps to see why Japan has suffered so badly from environmental damage during her intense development following the second world war.

Table 1. Comparison of demographic data in Japan, UK and the USA

	Population (millions)	Area (1000km2)	Agric. (Land Use %age)	Forest	Other	Pop.Density (per km2)
JAPAN	123.12	378	14.2	66.5	19.3	326
UK	57.20	245	75.7	9.5	14.8	233
USA	248.76	9,373	46.1	28.3	25.3	27

Source: U.N Monthly Bulletin of Statistics; Prime Minister's Office, Japan. 1989

From these figures it is easy to see that the population density in Japan is high. A more precise breakdown of land use reveals the desperate lack of space.

Table 2. Land use in Japan

Agriculture	14.4
Woodlands	66.9
Moors	0.7
Rivers	3.5
Roads	3.0
DWELLINGS	4.2
Other	7.4

Source: National Land Agency, Japan 1988

To see through all these statistics and to realize the degree of overcrowding in the Japanese urban environment, a more potent vision can be gained by observing the population density per habitable square kilometre. To couple this figure with the extent of the industrial output per habitable square kilometre, through GNP, it is evident that Japan's "economic miracle" has occurred at a high social and environmental cost.

Table 3.International comparison of population density

	JAPAN	GERMANY	UK	FRANCE	USA
Population per habitable km2	1,523	384	365	165	54
GNP per habitable km2 (US$million)	35.5	7.6	5.2	2.8	1.1

Source: National Land Agency, Japan

Land availability is not only restricted by the geographical conditions of Japan, but also by the land ownership structure and market conditions (Nakai 1988). In the urban development of Japan since 1945, the country has been transformed time and time again in response to the pressures of the market. Not only has the physical environment been transformed, but so has the social environment. Healey's (1992) "politics of turf" redefining suburbs has also occurred in Japan with many of the urban regeneration projects. Whole new areas have been created through large scale reclamation. The cities have been restructured and this has gone hand in hand with economic change and institutional restructuring. Developments in Japan have a multiplicity of actors but essentially, as in Britain, it is a partnership of state and the market. The position of environmental concerns within this changing frame of what Harvey (1985) calls "urban governance" has yet to be ascertained as the partnership between state and the market evolves

Post war development and its impact on environmental debate

The emergence of environmental debate in Japan has been the result of the severe impact of Japan's post war reconstruction and development upon its environment. Japan's environmental policy is integrally linked to its economic growth and development policy. To view Japan's environmental policy making in the post war period it is essential to analyse the stages of economic development and the reactive nature of government policy making. The post war period can be divided into four distinct periods which will be subsequently addressed by analysing the environmental policy making process in relation to economic growth.

a. Post war growth and industrial development (1955-1963)

b. Creation of serious pollution in the late 1960's and the rise of environmental consciousness (1964-1969)

c. Advent of environmental legislation and the OPEC oil shocks (1970-1979)

d. Economic recession, decreasing pollution and an emphasis on growth through public works. Bubble growth and burst

Post war growth

Following the war and the occupation period, Japan was faced with the unenviable task of re-establishing her economic base. The first National Five Year Economic Plan of 1955 pointed to growth and recovery through industrial development (particularly heavy industry), self sufficiency, foreign trade and an annual growth rate of 5 per cent. In 1960, development policies received a boost through Prime Minister Ikeda's "Income Doubling Plan". In short, this plan aimed to create "special industrial development regions" to redistribute both income and population. In 1962, as part of the First Comprehensive National Development Plan, fifteen cities were designated "New Industrial Cities" and two years later six areas were singled out to become "Special Areas for Industrial Consolidation". It is important to see that the Comprehensive Development Plan was not legally binding but was merely a guide. These economic plans worked well in the supportive working relationship with industry, unrestricted availability of reasonably priced petroleum and massive industrial growth. The GNP rose by an average of 10 per cent during the 1960s (Barrett & Therivel 1991).

The National Development Plans have tended to promote the "laissez faire" economic principles and government has done little to control business activities. The end of the war had reversed the government/business relationship. The defeat of Japan was a defeat for the government. It was the business sector that was to pave the way to Japan's recovery (McKean 1981). The resultant change in the government/business relationship saw companies no longer following state goals but initiating their own policies for development. From an environmental perspective the new policy makers were "stronger, more secure, more defiant, and perhaps less socially responsible" (ibid.)

Japan's reconstruction in the post war era

The major urban areas that had been allied bombing targets throughout the war had been decimated. The cities of Tokyo, Yokohama, Osaka, Nagoya, Kobe, Shimonoseki and Fukuoka were all reduced practically to ashes. The obliteration of first Hiroshima and then Nagasaki left Japan's major cities burning. The initial imperatives for the post war politicians working with MacArthur was to restructure Japan and to create a "democracy". The break up of the military and the zaibatsu was seen as fundamental. However as the "communist threat" emerged, containment became the predominant factor in U.S foreign policy. The "loss of China" and the subsequent Korean situation all helped Japan in her international rehabilitation. The rehousing of the Japanese people, of whom millions were displaced due to the war, took priority over planning the urban areas. Post war urbanization accelerated as people came to the cities in search of work. The post war economic recovery was given a boost in the early 1950s with industrial orders for the Korean war. This combination of circumstances left the urban areas in a vastly different form from their European counterparts who had gradually developed since the era of the Industrial Revolution.

The rapid urbanization of Japan has left its mark upon the urban environment. The traditional rural way of life, encompassing small agricultural hamlets, had developed so rapidly that it became a confused mass of enlarged villages coming together to comprise a town. Post war planning came about through governmental concerns for economic self sufficiency. Long term development strategies were set up based on regional development nuclei. The concept of the Pacific Belt emerged: a band of development to stretch from

Tokyo southwards through Nagoya, Osaka and on to Kitakyushu on the southern island. Regional differences in development were inevitable, especially when one considers the regional diversification within the Japanese archipelago. The problem of rehousing the Japanese population and rehabilitating her industries saw Japan embark upon a rapid development plan. Japanese development has, unlike that of many nations in Europe, been highly dependent on private financing for large scale construction projects. This has enabled the legacy of the pre war zaibatsu to re-emerge after MacArthur's purges, but in a different conglomeration.

Pollution and environmental consciousness

The success of the economic recovery was such that the Ten Year Economic Plan proposed by Prime Minister Ikeda bettered its own targets of growth. A peak of 14.5 per cent growth was recorded in 1961. The domestic market was benefitting from "income doubling". Mass production battled to keep up with mass consumption. In particular, the extensive industrial development programmes centred around the huge kombinato (industrial complex of manufacturing industry related to heavy industry) in decentralized areas, had created problems (Edmunds, 1983). As part of the rapid growth policy, these kombinato were located in densely populated areas to ease the burden of infrastructural improvements. These industrial complexes were generally located on the bay areas around Tokyo, Nagoya and Osaka. During this period of concentration on economic matters, the haphazard nature of these developments and their environmental impact was ignored. The strategy had "peppered the Japanese landscape with pollution, no longer concentrated primarily in major cities or large industrial zones as in other countries" (McKean 1981). The major side effects of this impressive economic growth were urban sprawl and pollution. Local governments depended on the redistribution of funding from central government for over half their funds. Local governments in areas of urban growth found themselves particularly dependent, having to provide increased infrastructure for the increased population. By attracting new industries, the local governments could increase their revenue. This gave rise to industrial growth in existing urban areas but also in new cities in the countryside. The pace of development was such that the problems created could not be handled by the conservative control at local level.

It was not until the late 1960s that voices started to be heard from those concerned about the environment in Japan. The "occupation authorities" had established tight political and social control upon the Japanese people through the "red purges". This strict regime was maintained throughout the

1950s. Any protest was harshly dealt with so protest remained minimal. To begin with the only protests seemed to come from those directly affected by the extensive pollution that was occurring all over urban Japan. Cases of poisoning from lead, cadmium and mercury were commonplace. The Japanese public were slow to react to the horrors that were occurring such as the deaths of seven hundred people killed at Minamata by legal mercury dumping into the Inland Sea. During the early 1960s, despite protests, the political system did not provide citizens with any effective institutional mechanisms with which to influence policy outcomes (Edmunds 1983). Japan became the "most polluted nation on earth, potential pollution was four times that of the USA and over sixty times the world average". Pollution continued at horrific levels into the 1970s. Diagram 1 indicates areas most affected by pollution in Japan in 1971. In the early 1970s public concern had reached such a level that the government was forced to act. By 1970 over 50 per cent of the Japanese were directly suffering from serious pollution damage (McKean 1981). Table 4 provides a summary of the main environmental problems in Japan since 1870 and the government reaction to date. The population had little recourse; those who disapproved "began to make demands on local governments that could not be handled within the framework of conservative consensus politics, so they formed citizens' movements instead" (ibid. p.25).

The Mishima 1963-64 revolt provided the government with a revolt which was ultimately successful in preventing a kombinato development. The domination of local government politics by local environmental issues became common and enabled citizens' movements to blossom. Pollution activists have a unique place in post war Japanese society. They have remained outside the system and yet have been successful: the system has been forced into accommodating them (Van Wolferen 1989). The Mishima defeat was a significant turning point as it alerted government officials to the rising resistance. In 1964, the government began to incorporate the citizens' concerns into development policy to avert any future protests (Edmunds 1983). It was also in 1964 that a Yokohama administration became the first government in Japan to obtain a pollution control agreement between city and power plant. This was a clear political success for the city's mayor and support was even echoed from MITI, keen to benefit from the enlightened policy making so as to prevent any further embarrassing project cancellations as in Mishima (ibid.).

The 1964 national election saw Eisaku Sato become Prime Minister in an election dominated by environmental problems. His platform was naturally that of fighting pollution problems. Central government funding for

119

Diagram 1. Pollution problems in Japan

JAPAN IN 1971

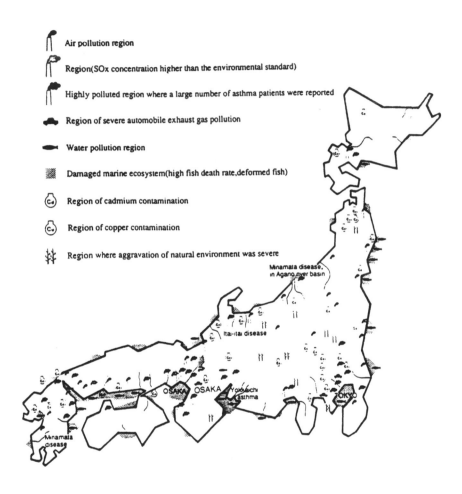

- Air pollution region
- Region(SOx concentration higher than the environmental standard)
- Highly polluted region where a large number of asthma patients were reported
- Region of severe automobile exhaust gas pollution
- Water pollution region
- Damaged marine ecosystem(high fish death rate,deformed fish)
- Region of cadmium contamination
- Region of copper contamination
- Region where aggravation of natural environment was severe

Minamata disease, in Agano river basin

Itai-itai disease

OSAKA OSAKA

Yokkaichi asthma

TOKYO

Minamata disease

Source: The Environment Agency - Japan

pollution increased rapidly and the Basic Law for the Environment was enacted in 1967. However, the law's "harmony clause" was ineffective as it lacked enforcement powers (Barrett & Therivel 1991). As the 1960s drew to a close pollution did not improve and citizen protest flourished across Japan. Finally a breakthrough in the citizens' use of litigation came to fruition.

Pollution related disease victims attempted to claim compensation but were frustrated by not only the companies concerned and the government, but also by their own communities. The breakthrough came in the form of the "Big Four" pollution cases of Minamata, Niigata, Tokkaichi and Toyama. The courts finally decided in favour of the pollution victims; much to the dismay of the defendants, the industrial conglomerates. This shift by the judiciary, as a result of citizen pressure coupled with the effective backing of the mass media, brought about a long overdue governmental reaction. Thus it could be deduced that the nature of environmental debate in Japan has changed and the focus in the 1990s is very different from those in previous decades. A basic breakdown of environmental concerns in Japan would include the categories in Table 5.

The advent of environmental legislation

The "Big Four" cases became an issue for national concern and hence for governmental reaction. They formed a "strong stimulus" for the subsequent environmental legislation of the early 1970s, including the "Polluter Pays Principle" which became the hallmark of this period. The cases had a particularly special significance in that they "conclusively demonstrated to a sceptical public the independence of the judiciary and the effectiveness of litigation in promoting justice" (McKean 1981). The handing down of the legal judgements coupled with popular discontent, hence political opposition, and worsening air pollution created a climate of change. The inevitability of change was clear for all to see; politicians and businessmen concluded action must be taken. It must be noted that governmental reaction in the form of the "Pollution Diet" was not the first governmental action to combat the problems. Action was taken on a local level where pollution is far more visible, those responsible politically more accountable and the reaction more flexible. The role of local government was instrumental in bringing environmental concerns into the policy making process.

Table 4. History of pollution problems and control measures

POLLUTION PROBLEM	DATE	COUNTERMEASURES
Discovery of health damage by mine pollution	1870	Prohibition of fishing in contaminated areas
Reported damage by sulphur dioxide	1880	
	1900	Forced evacuation from contaminated areas
	1910	
	1920	
	1930	
Reported damage by dust and soot pollution	1940	
Reported damage by heavy metals (outbreak of Minamata disease)	1950	Regulatory control on factory discharged waste water
		Financial subsidies for pollution control
Reported SOx damage (outbreak of Yokkaichi asthma)	1960	Regulatory control on factory exhaust gas Desulphurization technology development project Tax incentives to promote the implementation of pollution control equipment
Reported lead poisoning by leaded gasoline use	1970	Pollution related health damage compensation system
Reported scum problem		
Reported photochemical smog		Pollution Diet (Enactment of 14 laws concerning pollution control)
Outbreak of PCB contamination		Inauguration of the Environment Agency
Frequent reporting on noise pollution		United Nations Human Environment Conference
Reported eutrophication problem of lakes and ponds	1980	
Outbreak of high-tech industry related pollution		
Global environment problem becomes serious		
	1990	Conclusion of Ozone Layer Protection Treaty
		Holding of UNCED

Source: Environment Agency 1991.

Source: Environmental Agency 1991

Table 5. A typology of environmental debate in Japan in the 1990s

Energy Related Concerns:	Following the Oil Crises Japan's oil dependence was realized and alternatives investigated.
Mass Consumer Society:	The swallowing of all materials into the consumer led boom.
Natural Environment:	How to preserve Japan's natural beauty from rampant urbanization.
The Built Environment:	A 'softer' approach to the townscape and urban design, a harmonization of urban activities.

Local governments across Japan had reacted to protests and the highly visible effects of pollution. The national laws and guidelines had clearly not been effective and hence, faced with anti pollution activists, local governments had to solve the problems within the scope of their own powers (Tsuru & Weidner 1989). The cities that had experienced the most serious environmental problems due to their concentration of population and industry have tended to play major roles in guiding policy. Local legislation was considerably less stringent than national legislation. This process can be viewed in 1962 over a dispute over the legal validity of local ordinances over soot and smoke emissions being stricter than national law. The resolution of this conflict "created a principle whereby local ordinances may expand the scope of national laws and set stricter standards" (ibid).

National Policy making was stimulated by Prime Mininster Sato in July of 1970, perhaps coincidentally following a severe photochemical smog in Tokyo. The ministries were ordered to investigate pollution control measures and to strengthen environmental legislation. In February 1971 the cabinet proposed the establishment of an independent agency to coordinate the environment and pollution. In May of 1971 the Environment Agency was established by law and set up in July. Legislation ensued in the so called "Pollution Dict" of 1970-71. However, the majority of this "environmental legislation" was compensation for those whose lives had been ruined by the polluters. Thus the government showed a willingness to prevent pollution even though it risked economic growth. As a result, Japan has instituted one of the world's most comprehensive environmental legislative systems (Barrett and Therivel 1991).

The mainstay of the environmental legislation is the Basic Law For Environmental Pollution Control (1967). There are numerous other environmental laws as indicated by Table 6. As a result there was a marked improvement in air and water pollution (Diagram 2). Apart from the list of environmental laws in Table 6, there is a body of legislation relevant to development planning and land use control. Reviewing the list gives tha impression that they are mainly pollutant specific, focusing on industrial and commercial activities. However conservation and quality of life concerns in Japan received little attention at this time.

Diagram 2. Improvement of air and water pollution since 1971

Air pollution

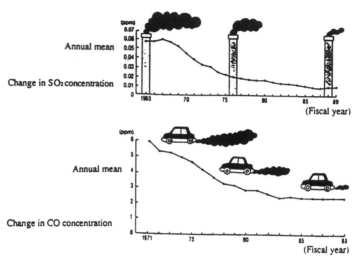

Annual mean

Change in SO₂ concentration

(Fiscal year)

Annual mean

Change in CO concentration

(Fiscal year)

Water pollution

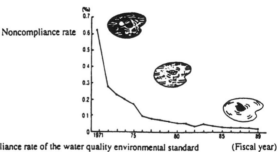

Noncompliance rate

Change in the noncompliance rate of the water quality environmental standard (health-related items)

(Fiscal year)

Source: The Environment Agency · Japan

124

Table 6. National laws on specific pollution

AIR POLLUTION
> Air Pollution Control Law (1968)
> Road Transport & Motor Vehicle Law (1951)
> Road Traffic Law (1960)
> Electric Power Industry Law (1964)
> Gas Industry Law (1954)

WATER POLLUTION
> Water Pollution Control Law (1970)
> Sewerage Law (1958)
> River Law (1964)
> Marine Pollution Control Law (1970)
> Hazardous Substances Control Law (1950)
> Agricultural Soil Pollution Prevention Law (1970)
> Seto Inland Sea Environment Conservation Law (1973)
> Regulation for waste treatment and disposal

SOIL POLLUTION
> Agricultural Soil Pollution Prevention Law (1970)

NOISE AND VIBRATION
> Noise Regulation Law (1968)
> Vibration Regulation Law (1976)
> Road Transport and Motor Vehicle Law (1951)
> Road Traffic Law (1960)

GROUND SUBSIDENCE
> Industrial Water Law (1956)
> Law Concerning the Pumping of Groundwater for Use
> in Buildings (1956)

OFFENSIVE ODOUR
> Offensive Odour Control Law (1971)
> Nature Conservation Act (1972)
> Pollution Related Health Damage Compensation Law (1973)
> Chemical Substances Control Law (1973)

NATURE CONSERVATION
> Natural Parks Law (1957)
> Nature Conservation Law (1972)
> Law concerning the Protection of Birds and Beasts and Hunting (1918)
> Law for the Regulation of Trade in Specified Birds (1972)
> Law for the Regulation of Trade in Endangered Species of Wildlife (1987)

Source: The new frontiers of Environmental Policy in Japan (Kazu, 1991)

Source: The new frontiers of Environmental Policy in Japan (Kazu, 1991)

Table 7. A chronology of events in Japan's environmental development

1947	Japan's Democratic Constitution enacted giving all the right to public participation and democratic freedoms.
1954	The Fukuryu Maru's (fishing boat) contamination by US nuclear test became the centre for an emergent peace movement.
1960	Evolution of the new-left, AMPO demonstrations against the revision of the Japan-US security treaty.
1963	The first defeat of a 'kombinato' development was achieved by the united action of the citizens of Mishima.
1967	Basic Law for Environmental Control passed.
1970	Environmental boom in Japan as residents rose up to oppose numerous development plans.
1970	The 'Environmental Diet' session.
1971	Basic Law for Environmental Control revised.
1971	Environment Agency established.
1972	Polluter Pays Principle (PPP) adopted.
1972	'The Big Four' cases established an environmental precedent.
1978	Narita Airport stormed by protesters at the height of a intense battle to prevent the development proceeding.
1980s	Increased environmental awareness of the Japanese people.
1993	Basic Environmental Law presented to the Diet.

Source: Authors

The activities of Japanese gooverment in tackling environmental problems were not only concentrated in Japan but also at the international level. In 1975 the Environment Commission of the OECD was invited by the Japanese government to examine Japan's environmental policies. There were also numerous "foreign aid" schemes set up to enable the efficient export of pollution. Most significantly, this period can be summed up as a success for central government. Environmental issues which had gripped the population in the previous decade had now been "addressed" and institutionalized. The problems of the economy now seemed more pressing. The system had responded to popular outcries amplified by a sympathetic press. Table 7 lists a chronology of environmental events. The introduction of strict industrial regulations reduced the pollution and calmed public opinion and indignation (Miyamoto 1983). It showed for all to see that governmental reaction could be brought about if public indignation is awakened but that it can only be brought about if the wrongs are sufficiently visible. In formulation of environmental legislation at this time there was no consideration to cost and benefit analysis. In the aftermath of the Minamata and Yokkaichi disasters, it was unthinkable for both Japanese public and private industries to consider the economic costs and benefits of any particular policy or legilative measure (Kazo 1991).

The oil shocks and pressure for change

When the first OPEC oil shock struck in 1973-4, Japan was 77 per cent dependent on oil for her primary energy. At that time Japan was 99.8 per cent dependent on imported oil, 85 per cent of which came from OPEC. The effect of the four fold increase in oil price between 1973 and 1979 was devastating to the Japanese economy which in 1974 recorded a GNP growth below zero (Barrett & Therivel 1991). Japan had to implement serious energy security measures to protect the economy as much as possible. The economic cost of the oil shocks and of implementing pollution control strategies are difficult to differentiate. Undoubtably the oil crisis hit many industrial sectors and had a detrimental effect on GNP growth but it appears that the anti pollution measures have not been entirely negative; not only have they helped the environmental situation but also they have been economically beneficial. Corwin (1980) observes that "the consensus seems to be that anti-pollution measures have, if anything, helped Japan's economic growth".

During this time Japan's energy security was of primary concern. Japan's phenomenal growth was threatened by spiralling production costs. In 1973 Japan was 77 per cent dependent on oil imports for total energy consumption. This crisis for Japan introduced high inflation which reached 26 per cent and sent the balance of payments into debt until 1976 (Reishauer 1988). The 1973 oil shock showed Japan how much of a security risk her energy dependency was; MITI's energy planners registered it as Japan's "Achilles Heel" and took measures to counteract it. Energy conservation, the use of more coal, hydro-electric power, natural gas and geothermal power sources were coupled with a boost in investment to promote the nuclear energy programme. The assumption that the price of oil and coal could only rise saw massive investments in nuclear energy for a cheap, reliable and independent energy for the future. By 1980 nuclear power generation had increased from 6,602 MW in 1974 to 15,511 MW in 1980 (OECD 1988). The scale of the investment was enormous: Japan invested nearly $1 billion a year on nuclear research, nearly three times that of the US nuclear industry (Asahi Evening News 21.7.91). These achieved targets must however be seen in the context of MITI's original plans. The targets for 1985 set in 1974 of 49,000 MW were a long way off. New targets set in the 1990 Energy White Paper were somewhat ambitious. It proposed the construction of forty more nuclear reactors of the 1 million kilowatt class by 2010 (Agency of Natural Resources and Energy 1991). By 1989 Japan had reduced her dependence on oil, imported from the politically volatile Gulf, for primary energy to a mere 22.1 per cent (OECD 1989). When the second oil crisis came in 1979 the effects were not as serious. Successful planning had removed a potential stranglehold on Japan's economy. Despite policy initiatives and investment from both government and business, Japan's energy efficiency remains poor. Tokyo alone in the summer months uses an equal amount of electricity as is consumed in the entire United Kingdom for the same time period (Nikkei Weekly 03-08-1991). Between 1973 and 1990, as a result of the introduction of energy conservation measures including good housekeeping, equipment improvement and process improvement, energy consumption per GNP declined substantially between 1973 and 1990 (Diagram 3).

Diagram 3. Trends of energy consumption per GNP since 1973

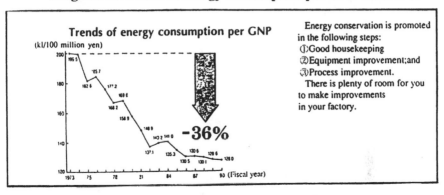

Note:GNP is based on 1985 price

Economic recession, decreasing pollution and an emphasis on growth through public works. Bubble growth and burst.

Following the oil shocks Japan's industrial structure experienced a serious restructuring. There was a decline in manufacturing industries and a shift towards the tertiary sector of services and information technology. Japan has exported much of her manufacturing industry to be nearer her foreign markets and competition from countries with lower labour costs have likewise taken the markets for many of the heavy industries for which Japan had been famous. This change is in itself a subject for discussion. Japan's recovery from a serious recession following the 1974 oil shock was remarkably quick compared with the west. This fact provides further evidence for those who point to the economic miracle and in particular the relationship between the state and the market. The result of the decline of manufacturing industry was a reduction in visible environmental problems of pollution. The current state of Japan's environment reflects the history of pollution problems as previously discussed. Economic policy in the 1980s has witnessed an emphasis on Japan's high-tech future. The development policy has had joint goals:

i.   the relocation of industry away from the developed centres bringing population and jobs to "under-developed areas';
ii. The development of the nation's infrastructure;

iii. Stimulation of the domestic economy

This economic policy has been warmly embraced by developers and the government. The public works projects

129

are a remarkably effective way of stimulating the economy but do have their drawbacks, notably environmental degradation. The sheer scale of the projects, the investment required, the environmental impact and behind the scenes political dealings make their construction of particular interest. Prime Minister Nakasone's influence upon this policy making was immense. The proliferation of public works went hand in hand with a stimulation of the housing market and a deregulation of the planning system through the relaxation of the Urbanization Control Areas as designated in the City Planning Act. Housing was seen as the key to stimulating domestic demand (Hebbert & Nakai 1988a). The Economic Planning Agency observed in 1982-3 "one may say that what holds the key to future expansion of the domestic market is whether or not qualitative improvement of the housing situation and urban redevelopment can be satisfactorily realised" (Economic Planning Agency 1983). The diversification of these projects is immense. Airports, shinkansen routes, expressways and land reclamation projects are amongst the most common projects underway across Japan at present. These development policies fuelled an explosion in land prices after a decade in which the rate of increase had steadily fallen (Hebbert & Nakai 1988b). The deregulation of the planning system was a major contributory factor to the growth of "the bubble" (Ghanbari Parsa et al 1993). The following section will identify the problems in Japan's environmental planning and policy making in the 1990s.

Problems of Japan's environment

The "environmental protection system" is limited and heavily biased towards pollution issues and the prevention of damage to human health. What the system does not emphasize is the protection of the natural environment and improvements in the quality of life. "To Japanese citizen activist and bureaucrat alike, the word "environment" is virtually synonymous with anti-pollution" (JEI, Nov 13, 1992). The policy making structure's treatment of environmental problems has been clearly shown to be reactionary. Pollution problems were "solved" through compensation and installing pollution prevention devices rather than by preventing pollution in the first instance. The emphasis on environmental policy has been through technological solutions rather than through implementing social change (Barrett & Therivel 1991).

Japan has been actively striving to protect the environment, promote health and safety, and use energy and resources more efficiently ever since pollution became a problem in the high-growth 1960's and especially since the two oil crisis of the 1970's, and now has some of the most advanced technologies and system in the world to reduce industrial pollution, enhance safety and hygiene, and conserve every energy and other resources (Keidanren 1991)

The above statement reflects the significant change in Japan's attitude towards environmental concern and sustainable development in recent years. This is partly to counter the international concern over Japan's actions over whaling and lack of concern for the tropical forests. The political perception of the domestic environmental situation has undergone great change as the Japanese public are exposed to increasing scientific evidence of pollution and destruction and press coverage of environmental issues, once a taboo subject in Japan. The population is becoming increasingly aware of these issues but it remains to be seen whether this concern will manifest itself into further political action on environmental matters as was seen in the late 1960s, or whether environmental concern will pass by as a fashion of the 1990s. The Keidanren's statement is expressing goals of sustainability but it perpetuates the concept of environmental improvement through technological improvements and growth. Japan's land use planning system is of primary importance in addressing the improvement of the broader environment.

The relationship between central and local governments

In understanding the complex interwoven relationships surrounding the incorporation of environmental factors into the development process, the relationship between central government and local government is vital. The governmental structures outlined in the following are crucial when considering the framework that surrounds any development and the legal imperatives involved in the approach to tackling environmental problems. The more complex a society becomes, the more the government needs to intervene to deal with the problems created by the complexities (Apter & Sawa 1984). The drawing of these jurisdictional boundaries by governments is critical to the success of the policies. In some cases, where the need for coordination and control takes precedence, governments become coercive and authoritarian to achieve their desired aims. In Japan the best example of this in recent times is that of the struggle of the Sanrizuka movement against the construction of Narita Airport.

It is therefore important to ascertain how the government decides on policy and who has a role in this policy making process. To understand the workings of the democratic state the mechanism of management must be clearly defined and this will be achieved by assessing the roles played by the different actors in the development process in Japan. There is a fundamental difference, when viewing "the system", on whether it is viewed from the top or the bottom. If the institutional mechanism of government fails to utilize and incorporate information from the bottom then extra institutional action can claim its position within the democratic process.

> By making accountability a function of direct opposition, the two become inseparable, a relationship converting democracy from a form of government in which political participation is passive, a form of complicity in power, into something dynamic, an expression of mutual responsiveness (Apter and Sawa 1984).

The position of such movements in the democratic system in Japan is undeniable. The failure of the ruling party and the ineffectiveness of the opposition parties failed to incorporate citizen dissatisfaction. Bokemann (1982) in his theory of "economic theory of democracy" sees the politicians as the corporate management unit directing the actions of the state. In the context of the Japanese situation this theory can be observed in the working relationship between the state and private business. The "extensive amount of collaboration in policy formulation" in this relationship is neatly illustrated by Eccleston

> Beyond the practice of amakudari (post-retirement employment of bureaucrats in big business) and such social practices, there is a long tradition of state involvement and a different perception to the state's legitimate right to guide the economy (Eccleston 1989).

In contrast, Eccleston refers back to the imposition of the 1947 constitution in Japan and the pre war continuous elements of the division of state and society. The function of the state is organically incorporated into the market mechanism of individual decisions (Kitamura 1976).

Environmental administration

The system of environmental administration in Japan is complex. Japan's highly centralized economy has strong legislative and financial powers with which to control the prefectures and municipalities. Japan's unitary structure consists of 47 prefectures and approximately 3,300 cities, towns and villages (Hidefumi Imura, in Tsuru & Wiedner 1989). In addition Japan's largest cities are represented by Metropolitan Governments. There are the twelve designated cities of Sapporo, Sendai, Chiba, Yokohama, Kawasaki, Nagoya, Kyoto, Osaka, Kobe, Hiroshima, Kita-Kyushu and Fukuoka as well as the Tokyo Metropolitan Government. As discussed previously, the post war reforms brought many changes to Japan's administrative structure. This was especially so in the reform of the local government administrative system which became more decentralized (ibid). The position of local government has been enhanced through the decentralization and administrative reform but, perhaps more significantly, through the Local Autonomy Act. This Act in principle empowers the local authority to manage public health, social welfare, education and economic matters on a regional scale.

Table 8 outlines the structure of national government, local government and the different actors involved in shaping the environmental policy in Japan. This melee of actors at both national and local level has combined to create the urban environment in Japan. Through their experiences of the severity of environmental pollution at first hand and the flexibility of local policy making, local governments have played a leading role in developing environmental policies in Japan. The establishment of the Environmenatl Administration in 1971 was followed in 1974 by setting up the National Land Agency, also within the Prime Minister's office. The Environmental Agency was responsible not only for pollution but also for the conservation of the natural environment (Japan, 1992, No. 533). At this time the need for controls was obvious although the technique of environmental planning was lacking. As Nakai observes "there was, however, virtually no land development control in Japan until the late 1960's" (Nakai 1988).

Table 8. Actors involved in the formulation of environmental policy

Government organisation	Function
Prime Minister's Office	Administration of work relating to minister's conference on global environmental conservation
Ministry of Foreign Affairs	Diplomatic policies and international cooperation
Ministry of Construction	Urban planning, construction of roads and dams etc
Ministry of International Trade and Industry	Research on environmenta conservation technology, industrial and trade policies etc.
Ministry of Finance	Fiscal, tax, monetary policies
Ministry of Health and Welfare	Dissemination of water supply, sanitary environment and food hygiene etc.
Ministry of Transport	Measures against traffic pollution caused by vehicles and aircraft noise
Environment Agency	Planning, designing and promoting basic policies concerning environmental protection as well as overall coordination within the government
National Land Agency	Planning, designing and promoting basic policies concerning national land use and coordination
Economic Planning Agency	Planning, designing and promoting basic economic policies and overall coordination
Ministry of Agriculture	Promotion of agricultural, forestry and fishery industry, management of national forests and authorization of agricultural chemicals
Ministry of Education	Environmental education in schools etc
Ministry of Home Affairs	Supervision by local and regional public authorities

Source: Environment Agency 1991, Environmental Protection Policy in Japan, Environment Information Centre.

With the establishment of these national agencies and the introduction of reactionary environmental and planning policies, the government set about creating a land use plan in an attempt to combat the problems created by rampant uncontrolled development. This rapid urbanization saw the corporations concentrate in the three main metropolitan areas to gain the cumulative benefits (Miyamoto 1991). The weakness of the Environmental Agency can clearly be seen in the attempts to introduce an Environmental Impact Assessment Bill (EIA). It was first presented in 1976 and was prevented from preceeding by the powerful ministries. The Environmental Agency has subsequently attempted to get the Bill passed yearly but with no success. Numerous actors intervened; commerce, industry and construction but especially the Keidanren (Woronoff 1986). The Environmental Agency finally withdrew the Bill in 1984 in the hope that some reforms would be introduced through some administrative guidance. The Environmental Agency introduced a "Basic Law on the Environment" in 1993 but again the opposition to a compulsory EIA led to the dropping of the EIA provision.

Conclusion

In terms of global environmental issues in the 1990s, Japan's experience has to form a crucial case study for the developing world. Clearly many environmental problems persist in Japan and are unlikely to be addressed with consequences that must be observed by those concerned with the advancement of sustainable development policies. In the context of Japan, it is evident that many of the improvements in environmental policy were brought about by social pressure expressed through citizens' movements, a political swing to the left at local level promoting environmental awareness and supported by the press. As pressure to adopt environmental protection measures subsides, so government and industry slacken their efforts (Tsuru & Wiedner 1989). In the 1990s, despite pursuing two decades of environmental policy, Japan's policy remains entirely reactive. This is not to say that there are no examples of effective environmental policy making. There is no theoretical concept for achieving a preventative environmental policy but the experiences of Japan can only prove useful. Likewise the role of those involved in the system can not be relied upon in the implementation of a preventative strategy (ibid.). Japan's "advanced technocratic environmental policy" is of benefit to the world and its worsening environmental problems in her investment into pollution prevention technology and information sharing. Tsuru & Wiedner (1989) point to three basic elements of importance in looking at Japan's environmental policy:

i.   Comprehensive problem centred environmental monitoring to enhance the comprehension of the issues;

ii. The effective opportunity of citizens to participate in policy making. This participation has broken down some of the "centralist elements" and introduced more stringent environmental policy than that required by law; an important step towards the dynamization of a sluggish political system with important lessons to be added to the international environmental debate;

iii. The legal leap forward enacted by the "Big Four" cases has introduced a fundamental dictum of democracy, legal parity, in this case between polluters and victims.

These elements are essential for the improvement of environmental policy. Despite the apparent potential for Japan to take a lead in environmental policy, the entrenched power structure has a crippling effect on policy initiatives (Janicke 1986). A reorganization of the system and the Environment Agency's plans for a Basic Environmental Law are currently being proposed as a big improvement upon the present situation, depending upon the success of the Environment Agency in getting the bill through the Diet. Regardless of the progress of the Bill through the Diet, currently in such deep political turmoil, the position of environmental concerns in the planning process appears secondary at best. The Bill has no provision to make a EIA statutory: hence there will be no compelling reason for the system of environmental management to be improved. On a practical level the current introduction of an EIA only in the final stage of planning looks set to remain for the foreseeable future. That is unless there is any conflict with primary goals such as economic or political imperatives but this seems somewhat unlikely. The power structure in Japan and the policy making process, although on the verge of "wide ranging reform" (Prime Minister Hosokawa 12-08-93), seems unlikely to address the broader environmental issues. Policy making looks set to continue hand in hand with economic imperatives, despite far reaching environmental initiatives proposed by the Environment Agency and others who appear to remain in a state of paralysis..

References

Apter, D. E. and Sawa, N. (1984), *Against the State. Politics and social protest in Japan*, Harvard University Press, Cambridge, Massachusetts.

Barrett, F.D. and Threivel, R. (1991), *Environmental Policy and Impact Assessment in Japan*, Routledge, London, p.16-26.

Eccleston, B. (1989), *State and society in post-war Japan*, Polity Press, Oxford.

Edmunds, C.M.W. (1983), *Citizen participation in a post industrial society. The case of Japan*, Unpublished PhD Thesis, University of Hawaii.

Ghanbari Parsa, A.R. and Penrice, W. (1992), "An examination of current issues in environmental planning and policy in Japan", paper presented at *Perspectives in the environment, interdisciplinary research on politics, planning, society and the environment*, ESRC Conference, Leeds University.

Harvey, D. and Denman, D.R. (1985), *Money, time, space, and the city*, Department of Land Economy, University of Cambridge.

Healey, P. (1992), *Rebuilding the city: property-led urban regeneration*, Spons, London.

Hebbert, M. and Nakai, N. (1988a), "Deregulation of Japanese planning in the Nakasone Era", *Town Planning Review*.

Hebbert, M. and Nakai, N. (1988b), *How Tokyo grows*, Suntory Toyota Centre, Japan.

Janicke, M. (1986), *Staatsversagen. Ohnmacht der politik in der Industriegesellschaft*, Piper, Munich and Zurich.

Japan Economic Institute (1992), "Environmental developments offer opportunities for Japan", *JEI Report*, January 10, 1992, Washington, D.C.

Kazu, K. (1991), *The new frontiers of environmental policy in Japan*.

Keidanren (1991), *Keidanren environmental charter*, Tokyo.

Kitamura, H. (1976), *Choices for the Japanese economy*, Royal Institute of International Affairs, London

McKean, M.A. (1981), *Environmental protest and citizen politics in Japan*, University of California Press, California.

Miyamoto, K. (1983), "Environmental problems and citizens' movements in Japan", *The Japan Foundation Newsletter*, 9(4), pp.1-12.

Miyamoto, K. (1991a), "Japanese environmental policies since World War Two", *Capitalism, Nature, Socialism*, Vol.2. (2) Issue 7, June.

Miyamoto, K. (1991b), *Regional Development*, Public Works and Environment.

Mouer, R. and Sugimoto, Y. (1990), *Images of Japanese Society, A study in the social construction of reality*, Paul Kegan International.

Nakai, N. (1988), "Urbanization Promotion and Control in Metropolitan Japan", *Planning Perspectives*, 3, pp.197-216.

Tsuru, S. and Wiedner, H. (1989), *Environmental policy in Japan* Sigma Publication.

Van Wolferen, K. (1989), *The enigma of Japanese power*, Macmillan, London.

Woronoff, J. (1986), *Politics the Japanese way*, Macmillan, London.

2 Can we have sustainability without the recoupment of development value?

Bob Evans

Introduction

In Western European societies, debates about sustainability are usually inextricably linked to questions of land management and use. It is clear that most, if not all policy prescriptions related to the achievement of some vision of a sustainable society are likely to have, often quite major, land use implications whether this concerns the protection of valued rural areas, waste reclamation or energy policy, or the move to a coordinated and more environmentally sound approach to transport. Land use policy is therefore likely to play a major part in any wider, integrated environmental policy which has sustainability as its ultimate aim.

In this chapter I wish to focus upon one aspect of land policy, the question of the ownership of development value, with a view to assessing the significance of this for the evolution of effective environmental policy. In particular, I wish to argue that, in Britain at least, public policies directed towards achieving sustainability are likely to be severely hampered, if not rendered completely ineffective, in many spheres, because of the absence of a mechanism for community recoupment of development value.

As will become apparent below, the debate over development value or the "unearned increment" has a long and hotly contested history. Moreover, the case for the public ownership and recoupment of development value has been put many times before, perhaps most convincingly by Eric Reade in what Peter Hall terms his "devastating criticism of the entire post-1947 British system of (land use) planning" (Reade,1987;Hall,1989,p192). However, there has as yet been little discussion of this issue in the context of policies for sustainability, the principal exception being the recent Town and Country Planning Association (TCPA) report on environmental planning (Blowers,1993). Although this report

emphasises the need for a betterment tax as an integral part of environmental planning for sustainability,
the justifications for this are unclear and the form that such a tax might take are not specified.

It seems to me that the TCPA may be correct in arguing for the recoupment of development value as an essential policy tool in the drive towards sustainability, but if this is to secure popular support, the reasoning behind it must be clear and concise, and the policy instruments specified. This is not a simple task and it requires considerable evaluation and investigation by those involved in environmental planning and policy. In the light of this, the following represents a review of some of the issues involved as a contribution to this important debate.

Development value and the "unearned increment"

There are two linked themes which run through the literature on development value. Firstly there is the matter of definition: what exactly are we talking about here?. Secondly, there is the question of ownership: to whom should this value belong? These themes, and the arguments which permeate and surround them have been extensively examined elsewhere (e.g. Clarke, 1965; Hallett, 1979; Reade, 1987; Cullingworth, 1980). My intention here is simply to provide a brief background in order to move to the wider question of land policy for sustainability.

What is development value?

Development value is most commonly understood as the difference between the existing use value of a piece of land and its value if it can be converted to a more profitable use. In contemporary Britain this usually refers to the increase in value of a piece of land which is brought about as a consequence of the granting of planning consent for a more profitable land use. Thus for example, the owner of farmland who secures planning permission to develop it for housing, or whose land is zoned for residential use, usually achieves a significant increase in the value of the land, and it is this difference between the existing use, agriculture, and the more profitable land use, residential, which is usually termed the development value.

In this case the increase in the value of the land has occurred because the "community", through the land use planning system, has decided that housing is a desirable land use. The increase in value has occurred through public action rather than the action of the land owner. There has been no change in the form or character of the land itself and for this reason, development value, following John Stuart Mill, is often termed the "unearned increment".

Clearly, land values can increase in other, somewhat similar ways. The overall level of economic activity may increase the demand for land or, as a result of infrastructural investment such as new roads or airports, land values in particular locations may increase. Nevertheless, in all of these cases the increase has been socially produced through the operation of market processes or by specific governmental action, and it is this which underpins the claim that since development value is produced by the community, it should therefore be owned by the community. I will return to this argument below.

In the case of state action through the land use planning process, which is the principle issue here, there is a further consideration. In the absence of planning restrictions, land owners could reasonably assume that if development in a particular area is likely to occur then it might settle on the land in their ownership. Clearly all land owners in the vicinity would be likely to have a similar level of expectation, but of course until it actually occurs, no one knows where the development will actually settle. This is usually referred to as the problem of "floating value".

This apparently irrelevant reflection only has significance if land use planning controls are introduced. When this happens certain areas are zoned for development and others are not. In this case the planning process has concentrated this floating value onto a few, specific sites. The value has been "shifted" and development value has been allocated to particular sites. The corollary of this, of course, is that the owners whose land has not been designated for development may feel aggrieved because they have effectively been prevented from realiasing any development value. An apparently simple resolution of this problem would be to tax those owners who had benefited from the shift in value brought about by planning, a tax on "betterment", whilst compensating those deprived of their development value. However, as Reade (1987) has pointed out this is a mistaken view, which we can now examine.

To whom should development value belong?

There are two possible answers to this question: the owner of the land; or "the state", representing the interests of the public or the community as a whole. If it is argued that development value is a socially created unearned increment which should be owned by the state, then certain policy actions flow. The development value should be recouped, perhaps through some kind of 100 per cent "betterment" taxation, and those land owners who are not allowed to develop should not be compensated because it is not logical to

141

compensate someone for the loss of something which they do not own.

Conversely, if it is argued that, in a free market economy, land owners should own development value, there should, of course be no taxation of betterment. On the other hand it would be logical to compensate those owners who were deprived of their development value as a consequence of public action through land use planning policy. Thus, as Reade quite rightly argues, it is logical to have *either* compensation *or* betterment but not both.

This was more or less the view taken by the post war Labour Government when it introduced that most radical piece of legislation, the 1947 Town and Country Planning Act. Informed by the conclusions of the Uthwatt Committee (1942) the Government took the view that it was necessary to nationalise both development *rights* and development *value*. Uthwatt had argued that the only logical resolution of the problems of shifting value and floating value set out above would be to enable shifts of value to operate within the same ownership. However, both Uthwatt and the Labour Government rejected land nationalization on the grounds of political controversy and administrative difficulty.

The 1947 Act was thus clearly predicated upon the belief that development value should be publicly owned, and therefore there would be no compensation apart from a short term transitional arrangement, but there would be a 100 per cent tax on betterment. This nationalization of development value was one of the few Labour post war policies in the sphere of urban and regional planning which was wholeheartedly opposed by the Conservative Party. This opposition combined with certain problems in the collection of the "development charge" and the peculiar nature of the immediate post war property market conspired to ensure that the Conservatives abolished the development charge in 1953.

The Labour Party attempted to revive the recoupment of development value on their return to office in the 1960s with the Land Commission Act, 1967 and again in the 1970s with the Development Land Tax Act,1976. In both cases their initiatives were repealed by subsequent Conservative administrations. Given this experience it is perhaps not surprising that recent Labour Party election manifestos have been largely silent on this issue.

Development value and land use planning

The Uthwatt Committee clearly recognized that if the new post war system of land use control was to be effective, it would be essential to link this with mechanisms for taxing betterment. Not only was this logical in terms of the arguments outlined above, but it was also necessary in order to ensure that the land use planning system had a sufficient level of control over the market so that it could be effective in actually promoting positive planning schemes rather than simply controlling development. The hope was that the 100 per cent betterment levy would eventually create a situation where all land was traded at existing use value, thus enabling public authorities to assemble land banks for development purposes with all development value reverting to the state.

However, with the abolition of the development charge in 1953 and a subsequent return to market value as the basis for all state compulsory land acquisition in 1959, the British land use planning system became what Reade terms "pseudo-planning, the appearance of planning without the reality. It seems likely that in such a system it will often be the market rather than planning which decides" (1987, p.23). The post 1953 land use planning system was both illogical in terms of dealing with development value, since there was neither compensation nor taxation of betterment, and largely impotent in the face of powerful property interests.

It might be objected that despite the abolition of betterment taxation, mechanisms do exist to enable community recoupment of development value. Firstly, Capital Gains Tax is currently payable on profits from all land sales, although of course development value cannot be recouped if the land is not sold, and like other general taxation measures it is susceptible to circumvention.

Secondly, local planning authorities and developers may enter into legally binding agreements under the 1990 Town and Country Planning Act whereby as a condition of the granting of planning consent, the developer undertakes to provide some service or development for public use. This might be the provision of a park, restoration of a church or the building of social housing either associated with the proposed development or elsewhere. This process is known as "planning gain" via "Section 106 agreements", and it is often argued that this represents community reclamation of the development value created as a consequence of the granting of planning permission.

The number of Section 106 agreements made have increased significantly during recent years, but it should be emphasized that these are *agreements* entered into voluntarily by prospective developers which are regulated by government circular and policy. It is not clear that such agreements are necessarily in the public interest, however that might be defined. Indeed, as Moore points out: "The giving of a public benefit (by developers) was often seen as a small price to pay in return for a grant of planning permission, and often enabled development which would otherwise have been controversial to be more readily accepted by the community" (1990, p.198).

143

There is little evidence to suggest that either planning gain or capital gains tax can be seen as effective measures to recoup development value. Equally they have not served to provide the land use planning system with greater influence over the land and property markets. On the contrary, as many commentators have pointed out (for example Ambrose, 1986, Balchin and Bull, 1987, Ward, 1993) the British land use planning system appears to be largely organized around the interests of the property industry.

Sustainability and land use policy

It is neither appropriate nor necessary to enter into a discussion here about the nature and character of sustainability (see for example Redclift,1987; Dobson,1990; Jacobs,1991; Agyeman and Evans,1994). For the purposes of this paper it is sufficient to recognize that, although sustainability is a contested concept, it is becoming a regarded by increasing numbers of people as a legitimate objective of public policy, although there might be doubts as to the current British government's level of commitment to the concept (HMSO,1994).

By the same token, I do not wish to review the increasing body of literature which focuses upon the importance of land use and planning policy as one element in a strategy for sustainability (see for example, Owens,1986; Elkin, McLaren & Hillman,1991; McLaren,1992; Blowers,1993;). Again, although there are differences of view as to what might constitute the most appropriate pattern of land use for a sustainable society, for the purposes of this paper I simply want to emphasize the significant role that land use policy must play within any overall strategy for securing sustainability.

Given these caveats, is it reasonable to assume that the current UK land use planning system can deliver the land use patterns appropriate for sustainability? In the light of the discussion above, I think that the answer must be "probably not". I have argued elsewhere that the town planning profession acts as a brake upon the development of polices appropriate for the "new environmental agenda" (Evans,1993). However, in the context of this discussion, the main problems to be overcome are the linked ones of, firstly, the comparative weakness of the land use planning system when confronted with the power of the market and, secondly, the effect of the current absence of policy with respect to development value which tends to stimulate development. I have already dealt with the first of these problems above, but the second requires some further elaboration.

Although much of the current discussion on sustainability and land use focuses upon the competing views of what might

144

be the "most appropriate" pattern of future land use, a linked question is that of the rate of current and future development, particularly in terms of the expansion of urban areas and the loss of agricultural and rural land. A recent Council for the Protection of Rural England report (Sinclair,1993) shows that the current rate of countryside "loss" to all urban uses is now about 25,000 acres per annum, more than twice the figure given in Department of the Environment returns.

I would suggest that a significant proportion of the pressure for this expansion is fuelled by the prospect of development value gains to be made from securing planning consent for change of use from agricultural to residential or commercial uses. In Britain, unlike some other countries, the property development industry does not usually distinguish between profits made from organizing the construction process and those resulting from the "unearned increment". There is therefore, a real incentive to pursue planning permissions on agricultural land, even if this may take many years.

For example, despite the current recession in the construction industry, there is evidence that some major house building companies are at present building land banks of sites in existing agricultural locations which are not currently zoned for development, or in some cases they are seeking options to purchase from agricultural land owners. The expectation, of course, is that planning permission will eventually be secured and that, if it is not, the land itself is a sound investment.

It might be argued that this is a reasonable and harmless activity, but as several authors have shown (for example Simmie,1985), such companies, either alone or through organizations such as the House Builders' Federation, have the capacity to exert considerable influence upon the decision making process. This may be either through corporatist influence in the plan making process or through high level pressure upon local authority officers and councillors. The point that I would wish to emphasize is that the prospect of achieving often quite substantial gains through the granting of planning consent ensures that there is likely to be, for the foreseeable future, a steady stream of land owners, their agents and developers, who have a clear interest in continuing to press for more and more development on land in "desirable" locations which currently have a low existing use value. There is little evidence that the current land use planning system has the capacity to resist this long term, influential and well informed pressure. It is also perhaps worth pointing out that it is not only large corporations who will have clear interests in maintaining this pressure. Virtually every owner of land with some kind of potential development value is a possible contributor to this process.

What are the options?

Before making some concluding comments it is necessary to briefly outline the various policy mechanisms which have been considered for the recoupment of development value. Aside from the options of capital gains tax and Section 106 agreements, which have tended in any event to apply only to larger planning schemes, and the politically unlikely solution of land nationalization, there appear to be three main policy contenders: a betterment tax; site value rating; and the auctioning of planning permissions.

Betterment tax

This was the arrangement incorporated within the 1947 Act, which was condemned as a tax on development by the then Conservative opposition, and was widely viewed as being as a brake on development. This was in part true, since in addition to removing an incentive to develop as no development value could be realized, owners in the post war period tended to delay development pending the promised repeal of the legislation. In current circumstances, and in the light of the discussion above, this tendency to act as a brake upon development might now be seen as a positive virtue.

Both the 1947 scheme and the later 1967 scheme adopted a betterment tax, although the tax was dropped from 100 per cent to 40 per cent. The later 1975 legislation also adopted a betterment tax but within a highly complex scheme which was widely viewed as unworkable (Boddy,1982). In all three cases the tax was portrayed by opponents as being over complex, bureaucratic and impractical, though of course, each scheme only lasted for a few years and was unable to develop the stability in operation that has emerged in other countries which tax betterment such as Sweden.

Site Value Rating (SVR)

A betterment tax applies only to land which is developed, and this means that owners of land who do not develop or sell for development are privileged in that they continue to benefit from increases in the existing use of their land, brought about by changing economic and market conditions. Thus they are still securing the unearned increment, and, like many owners in the post 1947 period, they can see advantages in withholding their land from development in that they can benefit from the increased value through rents or mortgages. This has led many observers to argue that *all* land value should be taxed, regardless of whether the land is developed or not. This tax would be fairer to landowners and, it is argued, would not reduce the supply of land for development,

146

since, in the case of land zoned for development, owners could hardly afford *not* to develop. SVR is, in effect, an annual levy on land which must be paid whether or not the site has increased in value.

Auctioning planning permissions

Peter Hall, like Eric Reade, argues that site value rating is probably the best way, theoretically, of recouping development value (Reade,1987; Hall,1989). Unlike Reade though, Hall takes the view that SVR is politically unattainable and instead makes the case for the auctioning of planning permissions (Hall,1989 p.191). Regional Structure Plans would allocate land uses in designated action areas in the inner city or greenfield sites, and the Treasury would then auction the development rights in these areas in the form of Development Bonds. The development value, or its approximate equivalent, would then return to the state in the form of the price paid at auction. Hall argues that this could be an effective and easily collected form of betterment taxation which has a reasonable chance of securing a measure of political support. "It might thus provide a way of ending the dreary game of political football that has run into massive extra time these past forty years, whereby Labour Governments pass laws to collect so-called betterment, and succeeding Conservative Governments promptly repeal them" (Hall,1989,p192). He continues later: "As Which? magazine might say, as compared with site value rating it might not be Best Buy, but it is Worth Looking At" (p196). This issue of what is politically feasible is obviously of crucial importance and will be returned to below.

Can we have sustainability without recoupment of development value?

I think that the short answer to this must be "no", for three main reasons.

i. The current land use planning system in Britain is mainly characterized by what has been termed "trend planning" (Brindley et al,1989). In other words, its predominant feature is the tendency to accommodate and support market trends. If sustainability is to be a real policy goal, a much more positive, proactive planning approach will be required which will inevitably need to offer strong opposition to market forces. As Reade quite rightly points out, this will only occur when taxation and land use policy are formulated together with the long term objective of removing the privileged position that the land market holds over the land use planning system.

ii. Sustainability implies a much more careful usage of all non-renewable resources, and a much more thoughtful consideration of how land is allocated and used than is currently the case. There is a real need to reduce the pressure for land development which is substantially fuelled by the opportunity that land owners have to make often quite large windfall gains of development value .

iii. The recoupment of betterment, even if this were to be done at a percentage considerably less than the 100 per cent tax of the 1947 Act, could provide a substantial income to fund public investment in environmental programmes.

To argue for the recoupment of development value in this way is not to ignore the other reforms to the land use planning system which will be needed if sustainability is to become a serious policy objective. In particular there will for example, need to be deprofessionalization of the land use planning process plus substantial democratization. However, the development value issue is clearly of great if not singular importance.

How should this unearned increment be collected: which is the best mechanism? The auctioning of planning permissions has the attraction of being the art of the possible, but the major disadvantage of being feasible only in certain, selected localities: action areas. It is difficult to see how such a system could operate nation wide. Site Value Rating on the other hand would have to operate nationally, and it is in many ways an attractive option. However, it does seem to be a particularly difficult policy to sell politically since it is, in effect a property tax with no immediate and obvious link to the process of development and the gains which flow from this. It has the makings of a deeply unpopular tax which all land owners would be compelled to pay, instead of falling just on those who chose to develop.

My own preference is for some kind of betterment tax on development, probably introduced gradually over a long period at an increasing percentage rate, eventually rising to 100 per cent. There would clearly be problems with such a tax, not least the question of valuation and the cost and difficulty of collection. On the positive side, it works in other countries, the tax discourages rather than encourages development and it has a simple logic related to a clearly unearned increment. There are also options for varying the percentage rate of taxation. It could be high in areas where development is to be discouraged and low or nonexistent in areas where the reverse is the case; furthermore, following the principle of subsidiarity, it would permit local, rather than central, control of taxation collection and expenditure.

However, regardless of the exact form of any betterment tax, it will be crucial to ensure that the proposed tax is integral to a wider environmental policy. The previous attempts to recoup development value were successfully opposed and repealed largely on the basis that they constituted a tax on development which was unjustified and unfair. Any new attempt must firstly be clearly based upon the legitimate right of the community to recoup betterment, and secondly must be seen as part of a wider process of strengthening the land use planning system in order to secure wider environmental policy objectives.

The problem here is of course, political. Many established, powerful interests and organizations within our society are implacably opposed to any scheme of betterment taxation, and their position is strengthened by the apparent total failure of the three schemes tried during the last fifty years. Moreover, the nature of the issue in hand is such that it will never attract widespread public passions. It is not a vote catcher and mobilization of popular support for this measure will be difficult if not impossible. Nevertheless if sustainability is to be any sort of reality, some kind of public recoupment of development value will have to be secured.

References

Agyeman, J. & Evans, B. (1994), "The New Environmental Agenda" in Agyeman, J. & Evans, B. (eds), *Local Environmental Policies and Strategies*, Longman, Harlow.

Ambrose, P. (1986), *Whatever Happened to Planning?*, Methuen, London.

Balchin, P.N. & Bull, G.H. (1987), *Regional and Urban Economics*, Harper and Row, London.

Blowers, A. (ed.) (1993), *Planning for a Sustainable Environment*, Earthscan, London.

Boddy, M. (1982), "Planning, Land Ownership and the State" in Paris, C. (ed.), *Critical Readings in Planning Theory*, Pergamon, Oxford.

Brindley, T., Rydin, Y. & Stoker, G. (1989), *Remaking Planning: The Politics of Urban Change in the Thatcher Years*, Unwin Hyman, London.

Clarke, P.H. (1965), "Site value rating and the recovery of betterment" in Hall,P. (ed.), *Land Values*, Sweet and Maxwell, London.

Cullingworth, J.B. (1980), *Environmental Planning 1939-1969 Volume IV: Land Values, Compensation and Betterment*, HMSO, London.

Cullingworth, J.B. (1989), *Town and Country Planning in Britain*, Unwin Hyman, London.

Dobson, A. (1990), *Green Political Thought*, Unwin Hyman, London.

Elkin, T., McLaren, D. & Hillman, M. (1991), *Reviving the City*, Friends of the Earth, London.

Evans, B. (1993), "Why we no longer need a town planning profession", *Planning Practice and Research*, Vol 8 No 1.

HMSO (1994), *Sustainable Development: The UK Strategy*, HMSO, London.

Hallett, G. (1979), *Urban Land Economics: Principles and Policy*, Macmillan, London.

Hall, P. (1989), *London 2001*, Unwin Hyman, London.

Jacobs, M. (1991), *The Green Economy*, Pluto, London.

Kivell, P. (1993), *Land and the City: Patterns and Processes of Urban Change*, Routledge, London.

McLaren, D. (1992), "London as Ecosystem" in Thornley, A. (ed.), *The Crisis of London*, Routledge, London.

Moore, V. (1990), *A Practical Approach to Planning Law*, Blackstone, London.

Owens, S. (1985), *Energy, Urban Form and Planning*, Pion, London.

Reade, E. (1987), *British Town and Country Planning*, Open University Press, Milton Keynes.

Redclift, M. (1987), *Sustainable Development: Exploring the Contradictions*, Methuen, London.

Simmie, J. (1985), "Corporatism and Planning" in Grant,W. (ed.), *The Political Economy of Corporatism*, Macmillan, London.

Sinclair, G. (1993), *Changes in Land Use 1945-90*, Council for the Protection of Rural England, London.

Uthwatt (1942), *Final Report of the Expert Committee on Compensation and Betterment*, HMSO, London.

Ward, C. (1993), "The deformation of planning", *Town and Country Planning*, Vol 62 No 7.

3 An assessment of the potential contribution of green belt planning towards sustainable urban development: A case in the West Midlands

Mark Baker

Introduction

This paper attempts to review critically Green Belt planning policy in the light of new commitments towards sustainable development. The focus for this examination will be on a case study of the West Midlands Green Belt with emphasis on a particular development conflict. This case casts light on many of the salient issues which now need to be confronted by policy makers in any reappraisal of the policy.

The Green Belt has become a focal point for many of the most bitterly fought planning conflicts in the region. There has been an absence of really definitive advice concerning the government's commitments to sustainable development. This has led to environmental NGOs, such as Friends of the Earth, taking a lead in trying to apply this concept to current land use planning conflicts, (Baker, 1992). The author was invited, with support from colleagues, to oppose a development proposal centering on the Green Belt and sustainable development. This was undertaken in the forum of a local public inquiry and a regional planning conference.

The paper highlights both the powerful link between road building and development and the weakness of planning constraints in the face generous road building programmes, particularly in Green Belt areas. The paper points out the conflict in a policy that attempts to regulate development whilst facilitating apparently demand led road building.

The Government has recently made commitments to the environment that could potentially have far reaching effects on existing policy in many areas. Many of these policies claim to have defence of the environment as one of their main components. To what extent does the objective of sustainable development compromise these policies? In addition, does the language with which these policies are currently expressed sufficiently encourage policy actors to address the new agenda in their decision making procedure? Many national and local government actors and NGOs from both industry and environment have adopted policies that support the concept of sustainability to some extent. Coming down from high policy making to application at the lowest level, however, there appears at first glance to be a large gap between policy and practice. Is it just a matter of time before the policy trickles down, or do far stronger institutional and political forces, that have so far escaped sufficient scrutiny, continue to impede or undermine any real changes?

In the last few years, definitions of sustainable development have been coming thick and fast from academics, environmentalists and consultancies alike. The evolving consensus appears to indicate a preference for a degree of compactness in settlement patterns to facilitate reduced energy use through greater use of public transport and combined heat and power schemes. Although the debate goes much wider than energy use there is not sufficient space to discuss it here. Good accounts are given elsewhere (for example Jacobs (1991) and (1993); Elkin et al (1991) and Breheny (1992)).

Introduction to the Green Belts

The emergence of the Green Belt policy, and the form it took, had much to with the way contemporaries viewed the environment in the early part of this century and the strength of various political lobbies. Therefore, access and amenity were key factors along with the general perception that the inner cities were unhealthy and polluted. It was considered to be for the general good of society that access to the countryside be secured for city dwellers by preventing urban sprawl. This was to be achieved by defining a wide belt of land surrounding major cities and conurbations within which development was to be tightly restricted. The policy was to be criticized in later years for being entirely negative. There was a large degree of ambiguity about the type of development that would be permitted.

Elson (1986), has demonstrated that the bitter political battle over the Green Belt policy in the early 1980s, encapsulated the Conservative government's struggle with local government and the desire for deregulation in the property market. Ironically, the result was that the Green Belt survived and may have been strengthened. Added to the early objectives were urban regeneration and the need to manage Green Belts positively. *Planning Policy Guidance Note 2* (PPG2) (DoE,1988) on Green Belts, reiterated much of the 1984 circular, emphasizing the policy's great importance and broad role. There were a number of subtle differences, however. When setting boundaries, land was to be allocated within the urban area to allow for long term development needs, then protection was to be for "as far as can be seen ahead". It is noticeable that in *Strategic Guidance,* (PPG10) (DoE, 1988) there is a commitment to monitoring "the success of the green belt in restricting the outward growth of the built-up areas and redirecting development to the inner city areas". This is of crucial importance for sustainability planning.

In 1988 the purposes of the Green Belts were defined as: to check the unrestricted sprawl of large built up areas; to safeguard the surrounding countryside from further encroachment; to prevent neighbouring towns from merging into one another; to preserve the special character of historic towns; and to assist in urban regeneration (DoE 1988). A report has recently been commissioned and completed for the government on the possibility of revision of the policy. This is in line with the rolling programme being undertaken on other policy guidance to incorporate the government's sustainability commitments. At the time of writing, this report has only just been published and there has been unsufficient time to incorprate its findings in detail (Elson, 1993).

The West Midlands Green Belt

Elson (1986) makes a number of interesting observations as to the original purpose set out for the West Midlands green belt. The proposal had been to mix decentralization and peripheral development. Birmingham City council was apparently in conflict with adjacent counties over its residential expansion plans being undertaken in association with inner city redevelopment. These became entangled in the "new town" debate. A major feature of the region was that industry had remained very centralized and therefore dispersal would necessarily mean greater commuting. In addition, population was continuing to expand up until the late 1960s. These circumstances led to a number of controversial exemptions to the green belt being granted. A much more positive signal was given in the early 1980s, with the development of green wedges primarily for recreational purposes, with active management. A further underlying feature of the region has been a decline of traditional industries and urban dereliction. This has led to pressure for diversification on new sites. Elson claims that the National Exhibition Centre is the most famous national example of exemption to the green belt. It was given as an isolated case. However, the Secretary of State for the Environment failed to acknowledge the associated development pressure that would ensue in the Meridan Gap. A noticeable factor has been the insecurity of the region economically. The planning authorities seem unable to resolve the conflict between the need for urban regeneration and the desire to release some greenfield sites. In this regard Elson accuses Birmingham of "marketing their green belt to cream off footloose industry that would otherwise go to the south east". In addition he concluded that "the influence of corporate power by housebuilders at the centre of decision making will continue to outweigh any minor victories obtained by outsider environmental interests". He asserted that abolition of the metropolitan authorities would strengthen the hand of developers, because local authorities would compete against each other. His criticisms presage many of those of Friends of the Earth in *Driven Out of Town*, mentioned below.

Towards a development consensus?

The current *Strategic Guidance for the West Midlands* (DoE, 1988) arose from the regional planning conference of 1987 and subsequent consultation. This was before sustainability commitments had been made by the government. It was issued in 1988, the same year as *Planning Policy Guide Note 2*. It appears to be full of contradictions. A heavy emphasis was laid on urban regeneration. In respect of Housing, the guidance talks of balancing with need to maximize house building in inner city areas with *"satisfying the demand for new housing outside the built up area"* (my emphasis). In addition it talks of the necessity of "high quality" development on the *periphery* "and this can be done without detracting from the commitment to urban regeneration". It notes however that they should not be used for retail or warehousing uses. Importantly, sites needed to have *"easy access to the motorway system"* "need not immediately adjoin the edge of the built-up area" and have *"generous parking"*. No concern here for reducing travel demand. All the strategic guidance notes have sections specifically on the Green Belt. In this guidance the Meridan Gap is singled out as being Green Belt that prevents neighbouring towns from merging. It continues to state that Unitary Development Plans should "take proper account of the likely scale and pattern of *development needs*" (my emphasis), and that "there is little evidence that a tightly drawn inner boundary of itself supports urban regeneration". The discretionary nature of the planning system is highlighted by the statement "it is unlikely that development for purposes other than high technology industry or housing will be permitted in the Green Belt". Housing represents the majority of development pressures and so this does not convey much reassurance in the integrity of the Green Belt and would seem unlikely to deter speculative applications claiming exemption. While the possibility of peripheral development remains, it does not appear impossible that development in more expensive and less attractive urban sites will be deterred to some extent. It is noticeable that local authorities were unable to gain private

funds to develop the Bescot Freight terminal in contrast to
Hams Hall, (WMPTSC, 1992). In relation to transport, the
priority is again stated to be urban regeneration. This is
qualified, however, by encouragement of "links into the
conurbation from the Birmingham Northern Relief Road and
the planned orbital route to the West of the conurbation".
There appears to be no acknowledgement of the likely
development pressure at those very access points which could
deflect development from the centre.

Map 1.

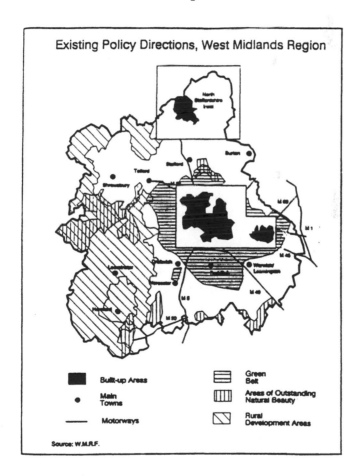

The government's position on the M25 sheds some light on
the Birmingham orbital situation. The then Secretary of State
(1987) stated that planners needed to have "positive attitudes
to the development pressures generated by the new

motorway". He also stated that the M25 does not affect adversely Green Belt protection. He even stated that improved recreational access to the Green Belt will be facilitated. No hint of sustainability here, in terms of traffic restraint and social equity. The inner city dwellers for which the Green Belt policy was originally intended to benefit, and who tend not to be car owners, would suffer.

The Europe 2000 report

The European Community is a highly dynamic entity whose influence is beginning to have some strong spatial influences on the distribution of economic activity, particularly in relation to achieving the Single Market. *Europe 2000* (COM 1991), indicates the peripherality of the West Midlands relative to the main growth sectors of the Community. In addition it seems to imply that development pressure in the region is likely to be most intense towards the east/south east of the region. In order for firms to be attracted to such regions and away from the centre of activity, they may be enticed by the provision of greenfield sites away from areas of high congestion. Current compensatory Structural Funds allied with domestic incentives may mitigate but not eradicate such pressures, particularly given the government's commitment to the role of the market. It is therefore not surprising that the Green Belt is coming under intense pressure in the West Midlands in general and in the east in particular. Planned road projects have been a major factor in this process.

The development culture - Birmingham's M25 - an NGO view

Between now and the year 2000, the Department of Transport has proposals which will amount to the construction of an outer orbital motorway system to surround the West Midlands conurbation. These schemes can be summarized as follows:

The Birmingham northern relief road (BNRR)
The western orbital motorway
M42 widening proposals
Associated feeder schemes

Friends of the Earth have conducted research, (FoE 1992), which indicates that elsewhere in the world where major cities have mature orbital motorway routes, major changes will have occured ino the economy, way of life, social structures and the environment of the existing urban areas. They believe that this would have very serious consequences for the long term viability of the West Midlands economy, and cause great damage to the local environment. The strong

158

development pressure now extant around the orbital routes illustrates the considerable dangers to urban regeneration policies. The government's rationale for these projects is to relieve pressure on the M6 and M5. FoE have challenged this, highlighting undesirable impacts of such extensive development proposals elsewhere.

The FoE prophesy has Birmingham and the Black Country losing premium industrial and commercial developments to the surrounding shire districts, thus reducing local government revenue for the urban area, which would be hit additionally by the drift of upper income groups into the countryside. Social dislocation would increase as the city becomes sharply delineated and abandoned to low income groups with few facilities. This view appears to be in sharp contrast to Planning Policy Guidance Note 10.

They believe also that far from relieving congestion, it will be made worse. The M25 experience and that of other orbital motorways indicate that local traffic and congestion will continue to grow, on both motorways and connected urban roads. M6 and M5 congestion occurs chiefly at peak times. FoE have shown that figures from the Department of Transport and the West Midlands Authorities suggest that the new motorway proposals will not significantly reduce traffic congestion on the M6 and M5.

The motorways will contribute to national traffic growth by providing significantly more road space and incentive to use cars despite the commitment to stabilize CO_2 emissions, and is therefore contrary to sustainable development objectives.

Development pressure reappraisal

Making the right choices, (WMRPFA, Oct. 1992)

In advance of a major regional conference aimed at revising strategic guidance for the West Midlands, a number of consultation reports were published. These were compiled by the West Midlands Forum of Local Authorities, to highlight potential regional development options of the future. "Making the Right Choices" was the general report which had followed earlier consulation over "asking the right questions". The latter, formed the basis around which the conference was based. It was noticeable in having a lot to say about sustainable development in general but not specifically in relation to the Green Belt. In relation to the Green Belt the report, under the heading "Choices in Green Belt and the Location of New Development" (WMRFLA, Oct. 1992, p.80), the challenge to the policy is set out. The challenge is how best to sustain the five principle functions of the Green belt whilst accommodating the development implications of meeting the region's social and economic aspirations, and how best to value Green Belt and green wedges as environmental, recreational and leisure assets.

The report compares a number of approaches which represent shifts in priorities in different areas of policy towards economic, social or environmental priorities and describes what these would entail in terms of policy.

Map 2.

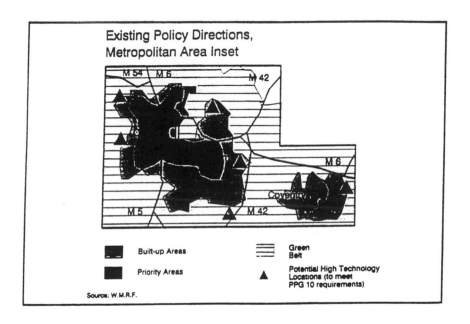

Major Green Belt issues in regional conference (WMRPF, 1993)

A whole day of a two week conference was devoted to issues involving Green Belt policy. Fundamental to this debate was an examination of whether the participating parties continue to support the policy as currently implemented. Various alternative growth scenarios were presented.

There was much support at the conference for a selective review of Green belt boundaries but only two advocated a fundamental review. Professor Cherry, in a paper which was written up to 18 months previously, did not address the issue of sustainable development, (Cherry, 1992). His advocacy was for a two tier green belt. The other advocate was the

House Builders Federation, justifying this position by the need to consider development needs up to 2021, (HFB, 1993).

The lack of sympathy for these positions was due to most contributors prioritizing urban regeneration, which they believed would be undermined by large scale land releases. Some expressed an interest in moving the inner boundary outwards to allow for some peripheral development and the idea of public transport corridors was widely advocated. Others maintained firm support for existing policy with overspill development to go beyond the Green Belt. The issue of commuting across the green belt was addressed by some as the reason for their support for peripheral development using public transport corridors and existing infrastructure. There were numerous contradictions in the presentations. Only one contributor addressed the conflict between achieving urban regeneration objectives while avoiding town cramming. One of the most radical local authorities was Canock Chase, (CCDC,1993), whic in a long presentation advocated a long term approach to urban regeneration. It believed that derelict land should be added to green space by reclamation, allowing for the creation of a better environment. In the longer term this would encourage reinvestment whilst in the meantime concentrating growth along transport corridors around redundant facilities. Cannock Chase recommended compensation by way of extending the outer boundary of the Green Belt. Very few addressed the orbital and those local authorities that did were in favour. Only Friends of the Earth, Midlands Amenity Societies Association and West Midlands Churches, (1993), pointed out the contradictions and dangers of the orbital road to policy objectives. Many of the participants believed that they could not oppose government policy with regard to the orbital road, but this was denied by the conference chairman, who indicated that it was their role to give advice that may alter government policy.

ECOTEC and This Common Inheritance

The government recently commissioned research on land use planning and sustainablity from the consultancy ECOTEC. At the time of the conference the full report had not been published, but it had been alluded to in "This Common Inheritance" second year report. It was particularly important because most of the study was based on Birmingham. It concluded that many existing policies based on prioritizing urban regeneration should be continued and extended and indicated that centrality and accessibility of employment to major public transport nodes was very important in reducing CO_2 emissions. The findings suggested that current policies in *Planning Policy Guidance Note 2* and *Planning Policy*

161

Guidance Note 10 should be continued (Bozeat et. al. 1992). No other participant appeared to be aware of the report.

Intervention to protect the environment

The Green Belt is the only policy concerned with restricting urban sprawl and preventing towns from merging. However the Green Belt in some areas does not appear to serve this function whereas in other areas such as the Meridan Gap, between Birmingham and Coventry, this function is seen as critical.

There are a number of measures to preserve historic towns, principally through the designation of conservation areas and general land use zoning. However, the vulnerability of such towns is clearly illustrated by the case of Coleshill, which sits uncomfortably close to the site of a proposed business park, the Hams Hall freight terminal and the Birmingham northern relief road. The latter stands to infill the small area still separating Coleshill from the Birmingham conurbation as can be seen from the map.

Map 3.

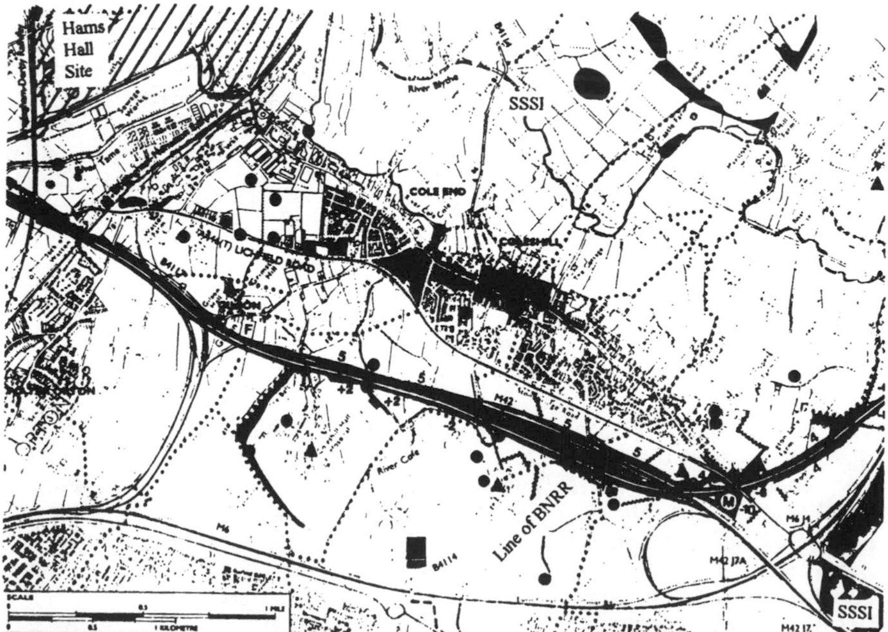

Projected route of the BNRR - Water Orton to the M6 River Blythe. Shows the proximity of Hams Hall and the BNRR to Coleshill and two SSSIs. To the South it is possible to see the previous encroachment of Chelmsley Wood estate into the Green Belt. Source: Adapted from BNRR Environmental Statement (Non Technical summary).

Evidence presented to the Hams Hall inquiry

In March 1992 the relatively recently privatized PowerGen closed down its last coal fired power station on the large Hams Hall site in North Warwickshire. There is not space here to look into the privatization and energy policy issues underlying the decision to close down the station and the alternatives although they underlie the whole issue of land use on the site. In July 1992 a planning application was submitted to North Warwickshire District Council on behalf of PowerGen and Trafalgar House Business Parks. It was perhaps inevitable that PowerGen would seek to exploit its land holding in association with an experienced developer. The application was to develop a freight terminal and manufacturing park on the power station site, while reserving sufficient land for a possible future gas fired power station. The site was in the Green Belt, but had received designation long after the power station was built. In addition the application had not been anticipated in the statutory planning process and contravened many of its policies. In association with the planning application, an environmental statement was produced and offered for consultation. The type of development proposed did not require such a mandatory statement but the local authority requested one under Schedule 2 of the regulations. In addition to the above, the sheer scale of the proposal in terms of site size, employment and its potential overall impact on the immediate area and the region, made it a certain candidate to be called in by the Secretary of State. The crucial aspect however was the Green Belt issue. It could not be claimed to be a development appropriate to a rural area and would therefore require exemption to be granted. In this respect a key area of contention was whether there was an alternative site.

A local public inquiry was held at the site of the former power station between 21st November, 1992 and 9th January, 1993. At the inquiry, the developers built their case around the environmental statement, which appeared to take on a much wider cost/benefit function. Like many such EIAs it could not be described as an objective document. Although it was very substantial in length, its coverage of some issues supportive of the proposal was much stronger than others.

The applicants stated their main case for the site to be

its strategic location in relation to the national motorway network, accessed within 1.5Km of the proposed development area, its existing rail connections, a size large enough to provide scope for ancillary distribution and manufacturing uses at a scale commensurate with the likely throughput. In addition, the regional and national importance of the Rail Terminal, its ability, because of its location, to serve both the East and the West Midlands and because of its past use, rendering the site derelict. (EAG, 1992)

It is a clearly defined site that can allow for an integrated freight village; the applicants placed great emphasis on this during the inquiry. They suggested that impact on the wider Green Belt would be reduced by landscaping and that the Green Belt function would be enhanced by increasing access and management to an adjoining nature reserve. The traffic impacts and additional infrastructure are the areas where the statement was most vulnerable. The apparent lack of balance in the statement can be seen by the claim made for air emissions reductions, as a result of the transfer of freight from road to rail, whilst ignoring the impact of emissions from employees and site visitors. The conclusion of the statement was that the economic benefits both locally and nationally would override the adverse environmental impacts. The application was supported by North Warwickshire District Council and Warwickshire County Council.

The main case against the application was made by the West Midlands Planning and Transportation Sub-Committee, (Wenban-Smith, 1992). This is a residual body from the old West Midlands Metropolitan Authority which attempts to maintain planning coordination for the conurbation. This body objected on two main fronts. The first was that the proposal was in contravention of the entire planning process both locally and regionally. The focus of conflict revolved around PPG2 and PPG10 and aspects of the development plans, the latter involving a large scale breach of the county structure plan in the allocation of employment land as well as the retrospective attempt of the local authority to incorporate the proposal within their development plan by amending the boundaries of the Green Belt to exclude the site (NWDC, 1993). The second plank of opposition lay in its support for an alternative site at Bescot. This lies within the conurbation. This had gone through formal approval in the planning process but had not been confirmed by the Secretary of State.

The case was complicated by the position of British Rail (Railfreight), who were rendered apparently impotent by imminent privatization and refused to state a preference for either site. Indeed it offered a third option, through

Map 4.

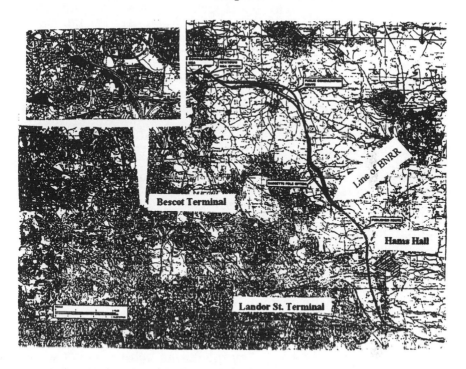

Projected route of the BNRR. Shows proximity to Hams Hall and to the Bescot freight terminal, with associated development sites (INSET). Source; Adapted from BNRR Environmental Statement (Non Technical Summary).

the media, of an interim expansion of an existing terminal at Landor Street in central Birmingham, apparently removing the need for a hasty decision (Crane, 1993).

The lack of an overall authority to take such decisions had greatly delayed and weakened the case for Bescot. The site fell on the edge of two local authority jurisdictions and was complicated by other planning applications that were in the process of consideration. Inevitably, in a more central area, space and access was much more limited and involved driving an access road through a school playing field, (Hurley, 1992) and using a small area of green wedge land. The main site consisted of 30 ha compared with 194 ha at Hams Hall. In addition, sites for associated development were dispersed. Whilst admitting the difficulties, the West Midlands Sub-Committee argued that the integrated freight village concept was not the only option, that the government had expressed commitment to urban regeneration, and therefore should work to overcome the difficulties: the Bescot site was much better served by public transport. It also argued that unemployment in the Bescot area stood at 18 per cent compared with 12 per cent near Hams Hall.

Some local authorities attempted to bargain with the applicants over improved provision of amenity and the building of a passenger train station at nearby Coleshill. The local M.P., Mike O'Brian, appeared in support of the proposal while at the same time opposing the Birmingham northern relief road. It was clearly admitted by the applicants that only phase 1 (about 0.33 per cent) could go ahead without the Birmingham northern relief road, which therefore appeared to prejudice the future inquiry on that project. This suspicion was heightened because of the involvement of Trafalgar House in both projects.

It was the link with the Birmingham northern relief road that aroused the objections of many NGOs and amenity groups, because of the considerable development pressure and previous Green Belt encroachments in the area. The objectors included Friends of the Earth, Transport 2000, the Council for the Protection of Rural England, local amenity societies and local residents.

If one removes the Birmingham northern relief road issue, the case seems a good example of the conflict between defending the Green Belt versus urban regeneration or town cramming. The delay in decision at the time of writing may indicate the considerable policy dilemmas behind the case. If the government approves the site it would appear to discredit its commitment to urban regeneration in contravention not only to those planning authorities but also the Black Country Development Corporation. If it refuses permission, it would probably be obliged to commit considerable resources to an alternative terminal, at a time of financial cut backs. The position has recently been complicated further by Railfreight who have stated that they no longer believe that the Bescot site is big enough for associated development.

Conclusion

There needs to be research and debate to ascertain what is a desirable level of urban greenspace. Once a target is established, it becomes easier to establish urban regeneration objectives and assess how far, if at all, the urban area needs to expand at the periphery. Such research would need to assess the requirements of developers but, in addition, should target particular urban sites for types of development which most accord with the attainment of a sustainable city. This needs to include further work on urban form to ascertain acceptable sizes of towns and cities for sustainability. There should be a thorough survey of every major town and city, with the objective of setting sustainability targets within develoment plans. This would have an impact on Green Belts within those plans.

More justification needs to be made within expressions of sustainability in order to prevent neighbouring towns from merging. This could be linked to the work above. When a target for such space has been established this should be afforded absolute protection. A more critical look at the size of Green Belts ought to be undertaken to ensure that these objectives were being met in individual areas. In addition, the impact of the policy on the sustainability of areas beyond the Green Belt should be fully examined.

Speculative inapropriate development proposals should not be allowed in the Green Belt, irrespective of any claimed for regional or national importance. Such proposals should be subject to the full development process to ensure that all relevent interests can be represented before any exemption is considered. There should be further guidance as to what constitutes an historic town. Clearly, Cambridge and Oxford are obvious examples, but what of Coleshill and Tamworth? Both the latter have historic cores, but have been subject to considerable development pressure on their approaches. How historic does a town need to be to claim Green Belt protection and what form should this take? Further reassessments need to be made of the impact of Green Belt on other natural resources such as water catchments areas, expanses of water and mineral resources. Green Belt policies ought to have a view to such wider environmental interests. For example, the need to extract a mineral resource should be justified on sustainability grounds. This may conflict with aggregate extraction for road building or open cast coal mining.

While further research is being undertaken, I would advocate establishing additional criteria for the purposes of Green Belts. This may involve the establishment of new Green Belts. These additional criteria would be to shape urban growth into more energy efficient patterns and to manage positively Green Belts towards sustainable development objectives. The latter would involve the establishment of initiatives such as the community forests and country parks as CO_2 sinks and deflectors of recreation from less sustainable sites. This would relax pressure on sensitive ecological sites and reduce the need to travel. A crucial criterion for the establishment of such sites would be accessibility. Sufficient resources would need to be made available for ongoing management and to realise the full educational potential of such projects. In establishing sites and management regimes there should be full participation given to the public. It could be argued that the latter should be given greater weight in less ecologically sensitive areas within the Green Belt. This may allow for far more restrictive policies to be adopted in areas of greater sensitivity.

Ultimately there needs to be a thorough review of the whole town and country planning system to see whether it can improve the delivery of sustainability. It can often be more difficult to modify an existing system than create an entirely new one, particularly where there are strong vested interests. The current system may continue to frustrate more radical efforts to move towards sustainability. This can be evidenced by the difficulties of effective environmental assessment and of attempts to introduce strategic environmental assessment.

Postscript

The Inspector, in his final report, gave his approval for the project subject to the conditions already outlined. He indicated some sympathy with the views of Friends of the Earth and other objectors but believed that the economic case overroad environmental objections. He avoided any reference to sustainability and expressed his views in a very traditional manner. The wider strategic impact on the environment from associated projects was ignored. Therefore the recent revisions of the planning guidance have barely penetrated the decision making process at the present time.

References

Baker, M.(1992), *Proofs of Evidence Submitted to the Hams Hall Local Public Inquiry by Friends of the Earth, Birmingham, Ltd.*, Department of Environment, Midlands Office, Birmingham.

Bannister, D. (1992), "Energy Use, Transport and Sustainable Settlement Patterns?" in Brehany, M. (ed.) *Sustainable Development and Urban Form,*   Pion, London.

Bozeat, George, Nick Barret, Jones, Glynn (1992), "The Potential Contribution of Planning to Reducing Traffic Demand?", *PTRC Summer Annual Meeting*, London.

Brehany, M. (1992), "Contradictions of the Compact City", a review in Brehany, M. (ed.), *Sustainable Development and Urban Form*, Pion, London.

Breheny, M. (ed.) (1992), *Sustainable Development and Urban Form*, Pion, London.

Canock Chase District Council (1993), Submissions to the West Midlands Regional Planning Conference, 16th

February "The Green Belt and Location of Development", *West Midlands Regional Planning Forum (WMRPF)*, Birmingham.

Cherry, G. (1993), "Green Belt and the Emergent City", *West Midlands Regional Planning Forum*, Birmingham.

Crane, D. (8.6.1993), pers. comm. *Railfreight Distribution*, Birmingham.

Department of the Environment (1988a), "Planning Policy Guidance Note 2: Green Belts", *Encyclopedia of Planning, Law and Practice*, Sweet and Maxwell, London.

Department of the Environment (1988b), "Planning Policy Guidance Note 10: Strategic Guidance for the West Midlands", *Encyclopedia of Planning, Law and Practice*, Sweet and Maxwell, London.

Elkin, T., Mclaren, D. and Hillman, M. (1991), *Reviving the City: Towards Sustainable Urban Development*, Friends of the Earth, London.

Elson, M. (1986), *Green Belts; Conflict resolution on the Urban Fringe*, HMSO, London.

Elson, M. (1993), *The Effectiveness of Green Belts - A Report for the Department of Environment*, HMSO, London.

Environmental Assessment Group Ltd. (1992), *Hams Hall Environmental Statement Vol.ii*, PowerGen & Trafalgar House ,Development Holdings Ltd., London.

Friends of the Earth (Birmingham) Ltd, Submission to WMRPF, Birmingham.

House Builders Federation (1993), Submission to WMRPF, Birmingham.

Hurley, P. (October - November 1992), *Proofs of evidence submitted to the Hams Hall Local Public Inquiry*, West Midlands Planning and Transportation Sub-Committee, (WMPTSC), Department of the Environment, Midlands Office, Birmingham.

Jacobs, M. (1991), *The Green Economy Environment, Sustainable Development and the Politics of the Future*, Pluto Press, London.

Jacobs, M. (1993), "Implementing Sustainable Development: The Practical Steps", *RSPB National Planners Conference; Planning for sustainable land-use*, RSPB, Sheffield.

North Warickshire District Council (1993), *Local Plan - Deposit*, NWDC, Atherston.

Wenban-Smith, A. (October - November 1992), *Proofs of Evidence Submitted to the Hams Hall Local Public Inquiry by WMPTSC*, Department of Environment, Midlands Office, Birmingham.

WMRPF (1992), *Making the Right Choices: A Regional Planning Guidance Consultation Report*, No.2, WMRPF, Shrewsbury.

West Midlands Amenity Societies Association (1993), *Submission to WMRPF*, WMRPF, Birmingham.

West Midlands Churches (1993), *Submission to WMRPF*, WMRPF, Birmingham.

West Midlands Friends of the Earth (1992), *Driven Out of Town? The West Midlands Motorway Proposals: The Impact on Economic Development and Urban Regeneration in the West Midlands Conurbation*, WMFoE, Birmingham.

West Midlands Planning and Transportation Sub-Committee, (WMPTSC) (1992), *Bescot a Freight Centre for Europe - The Prospectus*, WMPTSC, Birmingham.

4 An innovative public participation methodology for local planning

Julia Meaton and Margaret Anderson

Introduction

Local Authorities are increasingly having to consider public opinion in and reaction to their planning decisions. The next decade will witness a growing demand for stronger community participation that results in citizens having a real say in what happens to their local areas. Agenda 21, a major outcome of the United Nations Conference on Environment and Development (Quarrie, 1992), states that by 1996, local authorities should have undertaken a consultative process with their populations in order to achieve a consensus on a Local Agenda 21. Agenda 21 suggests that this should be achieved through a consultative process during which the local authorities would learn from citizens and from local, civic, community, business and industrial organizations and acquire information needed for formulating the best strategies (Quarrie, 1992). This paper will briefly examine some of the problems of achieving this degree of public participation and will then describe the development and application of a new methodology that could counter many of these difficulties.

One of the fundamental problems of dealing with public participation is how to describe it. There is no clear definition and although it is generally agreed to be an important issue, interpretations can vary widely. As a result public participation exercises can be diverse, ranging from token consultation through to full scale citizen empowerment. In most incidences, however, it is normally closer to the former. The reasons for this are many, but can be placed into three categories: problems of community/plurality; apathy and uneven representation; and fear of losing control. For example, in any one community there may exist a number of different publics, often with different needs, aspirations and attitudes. These "publics" will frequently disagree resulting in conflict. In order to minimize this conflict, tokenism or consultation only with elite groups might occur, thus undermining the concept of true consultation (Crawley, 1992).

Many consultation exercises have been characterized by the under representation of one group and the over representation of another. Apathy, ignorance, lack of confidence and the belief that they cannot influence the outcome combine to prevent the majority from contributing to the debate, while on the other hand a minority with knowledge and confidence are often over represented. Planners and elected representatives can also compromise the consultation process with their fear of losing control and a belief that "they know best". The combination of these problems has often resulted in disillusion for all parties and a lack of confidence in the whole public participation process (Vining and Ebreo, 1991).

This paper will describe an experimental methodology which could be adapted and applied to a range of planning issues. The methodology aids the breakdown of many barriers to participation and citizen involvement and as a consequence could have an important role to play in the application of Local Agenda 21.

The Ashford Methodology

The Ashford methodology evolved from a project examining landscape change on the town's urban fringe and the public's awareness and preferences for these changes. The research wanted to explore the public's relationship with the countryside around the town and specifically to find out:

whether they visited it, whereabouts they went and what they did there;

whether they had noticed any changes and whether they liked or disliked them;

which areas they wanted protected and why.

Ashford is a growth area for Kent and as such has been targeted for a substantial amount of new residential and commercial/ industrial development in the next ten years. The study additionally wanted to find out how people would like to see Ashford growing in the future and we therefore wanted to ask questions about:

where the new residential developments should be located and what types of homes should be built;

where employment land should be located and what types of employment would be favoured;

172

where recreational facilities should be sited and what types of facility are favoured.

The survey also wanted to explore the public's thinking behind these stated preferences.

Initially it was planned to conduct a conventional questionnaire with a random sample of households in the Ashford area. However it soon became clear that such a tool would not be able to elicit the spatial and visual aspects of the survey. Nor would it allow the respondents to appreciate the full implications of their "planning" decisions. In order to achieve fuller interpretation of all the issues it was decided that visual aids such as maps and photographs were required.

The methodology that was finally developed arose out of two different approaches to environmental perception. These were a Japanese experiment called "Gulliver's Maps" (Nakamura, 1989) and the Yorkshire Dales landscape interpretation project "Landscapes for Tomorrow" (O'Riordan, 1992).

Both of these projects involved respondents giving and receiving information through the use of a visual aid. The "Gulliver's Maps" experiment employed a very large scale map laid out on the floor upon which local residents were encouraged to write down anything they wanted to say about any part of the area covered by the map. There was no focus for the participants who simply had instructions to express and record any feeling, idea, opinion or emotion they had about their local neighbourhood and environment. The researchers' aim was to build up an "image" of the respondents' environment based on the comments and expressions recorded on the map.

The Yorkshire Dales landscape survey used a floor game which represented a track through the Dales. Respondents following the track had to make a series of decisions about how they would treat various landscape features such as dry stone walls, meadows, and barns which consequently led to the landscape which would emerge as a result of these choices. This study was clearly much more focused than the Japanese project, but they both shared the idea of creating an image.

Both of these methodologies seemed to have relevance to the Ashford research. Drawing from both projects, the research team finally developed a public participation technique employing aerial photographs. Respondents were asked to annotate the photograph in answer to a number of questions, but were also encouraged to record anything else they thought was important.

173

The Methodology

The survey was conducted during one week in October 1991. Four aerial photographs, approximately 1 metre square, were stood on easels in a covered shopping mall in central Ashford. Each photograph was annotated with various place names and a "you are here" spot, to help respondents get their bearings. An exhibition of photographs showing landscape changes and possible future developments was also erected along with brief explanatory notes.

Each aerial photograph was staffed by an assistant whose duty it was to explain the aims of the survey and then to ask the questions and record any comments. When someone approached a photograph, they were told what the survey was about and asked if they wanted to participate. When they did they were handed a pack of coloured stickers and were asked to place them on the photograph in response to a series of questions. For each new participant, a fresh acetate sheet was placed over the photograph so that a complete record of each individual's choices was collected. The assistant read out the questions and also recorded any comments the participant made during the survey in order to gain greater insight into the reasons behind the respondent's strategy decisions. Initially it had been intended that the respondent would be left completely alone to read the questions, stick on the stickers and record their thought processes. However the pilot indicated that this was too cumbersome for participants, and also that they benefited from an assistant being able to orientate them and answer any queries.

The first few questions were relatively simple and asked people to place a sticker on their homes and on their workplace, if appropriate. The first was normally easy to do since one of the first things most people did on seeing the photographs was to identify and point out their own homes. More seriously, these early questions helped the respondents to orientate themselves, while at the same time giving us a clear indication of their everyday activity zones. The next questions asked where they went in the countryside, how frequently and what they did there; again building up a picture of their activity zones. Gradually the questions became more testing as the respondents were asked if they had noticed any changes in the countryside, what they were and whether or not they liked them. Respondents were then asked to place stickers on areas or places which they thought should be protected from change and development.

The most difficult questions concerned the targeting of land for residential and employment development. The respondents were reminded of Ashford's role as a growth town, and were then told to allocate the land for development. The stickers to be placed on the photograph had been scaled to represent exactly the amount of land required for this development. All of the stickers had to be placed on the photograph. This caused some considerable concern because of the very large area of land, and consequently large number of stickers, that needed to be allocated.

Following this very difficult question, the respondents were then asked to identify places for recreational activities and to decide what kind of facilities they should be. Next they were asked to identify areas they thought would benefit from more tree planting or more water features such as lakes and ponds. The three final questions were really just for fun with respondents being asked to place pink heart shaped stickers on places they really liked, and black spot stickers on places they really disliked. Finally they were given a gold sticker which respondents were invited to place where they thought there might be buried treasure. This last sticker was included as an incentive for people to participate as several modest prizes were on offer.

During the exercise, the assistant wrote onto the acetate any comments, emotions and ideas the participants had, building up a more complete picture of the reasons behind the decisions and choices displayed by the respondent. Although the assistant did not lead or advise the participants, they were able to answer specific questions and queries.

Once the participant had finished allocating the stickers they was asked questions in order to elicit background and socioeconomic information. These included the length of time they had lived in the area, whether they had ever participated in any consultations over the Local Plan or any other proposed changes in the area, and their job and age group. These answers were recorded on a separate questionnaire.

Results

This paper is predominantly concerned with the methodology developed for the Ashford survey, so the results discussed here will be limited to those particularly pertinent to that aspect of the project. For a more detailed account of the results see Anderson et al, 1994.

The aerial photographs were undoubtedly very successful in attracting people to participate in the survey. Few people were able to ignore the exhibition and even if they did not participate themselves, many took a keen interest in others who were doing it. During the 6 days of the survey, 288 acetates were completed although many more people actually participated in the exercise as some acetates were family and group efforts. Each acetate took between 20 and 40 minutes to complete, although some took longer with one woman taking over 90 minutes!

Although there had been some press releases and radio announcements concerning the survey, it seemed that most participants became interested as they walked past and had not been influenced by prior publicity. The inclusion of a competition with prizes was also intended to encourage participation, but it seems unlikely that such an incentive was really necessary.

The participants were entirely self selecting with no attempt to structure the response. However, Figures 1, 2 and 3 demonstrate that a good cross section was achieved.

Socioeconomic characteristics of participants in the Ashford planning survey

Table A Employment Status of Participants

EMPLOYMENT STATUS	Number	%
Unemployed	16	6
Housewife	29	11
Skilled	18	7
Semiskilled	21	8
Unskilled	21	20
Clerical/retail	53	8
Student/school	27	10
Retired	38	15
Professional/manager	45	17
TOTAL	268	100

Table B Age of Participants

AGE	Number	%
11 - 20	30	11
21 - 30	42	15
31 - 40	64	23
41 - 50	62	22
51 - 60	38	14
61 and over	41	15
TOTAL	277	100

Table C Gender of Participants

GENDER	Number	%
Male	130	47
Female	149	53
TOTAL	279	100

This strengthens the belief that this methodology is attractive and approachable for a wide spectrum of citizens and encourages all types of people to participate. These results are particularly interesting with regard to Agenda 21 which highlights the need to encourage more women and young people to participate in decision making and planning processes.

The first few questions concerning the location of peoples' homes, workplaces and countryside areas that they visited gave us a clear picture of the participants activity zones or their "everyday life landscape zones". One of our hypotheses was that people would want to protect these areas and would therefore display NIMBY (Not In My Backyard) tendencies when designating land to be protected. Although initially this seemed to be true, it became clear that this was largely because people wanted to protect the places that they were familiar with and these, apart from "honeypot sites" like Wye and Hothfield, tended to be places closer to respondents' homes.

The NIMBY hypothesis was further discounted when it came to the allocation of stickers representing housing and employment land. We had anticipated that respondents' local areas would be avoided when allocating land for development. When the respondents first realized the number of stickers that had to be placed on the acetate sheet, most responded with concern and alarm, with a minority refusing to allocate them all. However, most got over the shock and were able to continue. Most started off by placing them in the obvious places: for example, in the green areas closest to the town; but soon realized that in order to allocate all the stickers they would have to adopt a planning strategy. These strategies included rounding off the town, expanding villages, creating new settlements and various combinations of the three. Although there might have been NIMBY tendencies in the initial stages of allocating these stickers, respondents quickly recognized the need to make trade offs. This often resulted in them allocating land for development quite near and, in many cases, adjacent to their own homes. Areas that they had targeted as protected areas were also sacrificed for the "greater good" so that the adopted strategies could be achieved.

The discovery that most of the respondents were able voluntarily to accept such compromises is a major result and is an indication of the potential this type of methodology could have for use in local development issues. The acceptance of compromise came largely out of greater understanding. This suggests that the more aware members of the public are of the full reasons for various developments or site allocations, the less likely they are to respond with knee jerk reactions. Understanding will not necessarily make them approve of it, but they may be less antagonistic. In the Ashford study the full realization of the extent of the town's projected growth actually resulted in many expressions of sympathy for the planners: "I wouldn't want to be a planner" being a typical comment.

Just over a third, 34 per cent, of the respondents claimed to have looked at the draft Ashford plan. This surprisingly high number is probably due to the timing of the survey (in the waiting period between publication of the Deposit Draft in June and the Public Local Inquiry in December) and also local publicity about the choice of Ashford as a growth point and the town's relationship with the Channel Tunnel developments.

The fact that two thirds of all the respondents in the Ashford survey had not participated at any level concerning the Local Plan suggests that this methodology has the ability to encourage all types of people to take a greater interest in their local environment. It seems that many are also prepared to consider very carefully all the issues involved and welcomed the opportunity to register their often very strong feelings about the town and its future. The methodology also facilitates an exchange of information, since while allowing citizens to make comments and express their opinions, it also collects wide and often longstanding information about the town. Such information could be useful in explaining why particular areas are so sensitive and why people object to specific plans.

One of the most important findings to come out of this methodology must be its ability to attract people of all ages, abilities and backgrounds. Most public participation methodologies tend to be biased towards the articulate middle classes, with many professionals assuming the remainder to be unconcerned about proposed changes. This methodology has shown clearly that this is not the case, and suggests that many of these under represented people find conventional means of participation too daunting. The Ashford method proved to be fun, accessible, nonconfrontational and effective, with many respondents saying how much they had learned and how much they had enjoyed the experience.

178

Before hailing this method as the panacea to all participation exercises, it is clear that there are certain negative aspects. Firstly there is the problem of knowledge and awareness of an area. Many of the respondents in Ashford had a deep knowledge of the areas close to their homes and places they visited, but few had comprehensive knowledge of the whole of Ashford and the area covered by the Local Plan. This incomplete awareness obviously affected respondents' planning strategies. Similarly most respondents were not aware of any of the planning restraints and again this would have influenced their strategies. Some respondents had difficulty in understanding the aerial photograph and failed to differentiate between land uses. For example several interpreted all green and brown land as being available for development and did not differentiate between playing fields, woodland, crops, and town parks.

None of these shortcomings is unsurmountable. All that is required to counter them is an adoption of a more interactive dialogue between the planners and the public. Enhanced explanations and the identification of legal and physical planning restraints would solve most of these difficulties.

A further positive aspect of the Ashford methodology is its adaptability. The idea can be developed for projects on many different scales. The researchers are already preparing a number of future projects in liaison with the Ashford Borough Council and hope to modify and develop the methodology for a range of different applications. The method, for example, could be adapted for both a small site specific planning problem and a national or even international planning problem. On an international level it has other advantages as it is an ideal methodology for cross cultural comparative studies since the importance of language is minimized.

The Ashford methodology has emerged as an important technique for eliciting public opinion on planning and land use issues. It has potential widespread application and is a positive step towards achieving greater community involvement in a variety of fields.

References

Anderson, M., Meaton, J. and Potter, C. (1994), "Public Participation; an approach using aerial photographs at Ashford, Kent", *Town Planning Review*, 65 (1), pp.41-58.

Crawley, I. (1992), *Public Access to the Planning Service*, paper presented to the Royal Town Planning Institute (East of England) Conference on Participation in Planning, 11 May 1992 (unpublished).

Nakamura, M. (1989), *The New Phase of Participation and Gulliver Maps as a Tool*, Papers in City Planning, 23, City Planning Institute of Japan, Tokyo.

O'Riordan, T., Wood, C. and Shadrake, A. (1992), *Landscapes For Tomorrow: interpreting landscape futures in the Yorkshire dales National Park*, Grassington, Yorkshire Dales National Park Committee.

Quarrie, J. (1992), *Earth Summit 1992*, The Regency Press, London.

Vining, J. and Ebreo, A. (1991), "Are You Thinking What I Think You Are? A Study of Actual and Estimated Goal Priorities and Decision Preferences of Resource Managers, Environmentalists and the Public", *Society and National Resources*, 4 (2) pp.177-96.

Funding by The Leverhulme Trust for this research is gratefully acknowledged.

Part IV
STATE, THEORY AND POLICY

1 Towards a theory of the green state

John Barry

"We're all in the gutter, but some of us are looking up at the stars", Oscar Wilde.

Introduction

The characterization of green politics as either "anarchistic" or a modern form of anarchism has wide currency. Both within and without the green movement, its distinctiveness is held to reside in its embodiment of traditional anarchist values for modern ecological conditions. This self understanding is particularly evident in the almost complete monopolization of the green imagination by an anarchist vision of the society greens would like to create. The many pastoral utopias that litter green political literature, a society made up of small scale, face to face, decentralized ecologically sensitive communities, pay eloquent testimony to the common judgement that "greens are basically libertarians-cum-anarchists" (Goodin, 1992, p.152). While soviets plus electrification equalled socialism for Lenin, it seems that for many green theorists, activists, and commentators stateless, selfgoverning communities plus solar power equals the "sustainable society". No matter how beguiling this "small is beautiful" political theory is, one must always remember that beauty is largely in the eyes of the beholder. There are problems with eco anarchism, not least of which is that often greens are blind to the need for developing a critical political theory with reasoned principles as well as promises of a better future. The aim of this essay is not to banish the eco anarchist vision from the green pantheon, but to attempt a critical dialogue with it and to propose a more appropriate reintegration of its insights into green political theory.

Several reasons can be given for the negative attitude of greens to the state. First, the strong influence of anarchism on the development of green political theory underwrites a rejection of the state on the grounds that its existence is inextricably bound up with the ecological, political, social and ethical problems that greens are concerned with solving. The state is often regarded as the public embodiment of the "industrial paradigm", protecting an "environmentally hazardous dynamic" (Carter, 1993, pp.45-9), or the embodiment of "materialism, institutionalized violence, centralization, hierarchy", values and practices antithetical to the green perspective (Porritt, 1984, pp.216-17). Second, green suspicion of approaches that give a prominent place to formal political and legal institutions is tarnished by the authoritarian nature of previous "green" defences of the state, advanced by Ophuls (1977) and Heilbroner (1980) amongst others, which stand at odds with its libertarian and democratic self understanding. Third, more "benign" readings of the state's role in dealing with ecological problems are held to invariably favour technocratic as opposed to political solutions. In other words, only a "technocentric" outlook (O'Riordan, 1981, pp.11-19), premised on a crude instrumentalist valuation of nature, is compatible with the bureaucratic administrative logic of state agencies. Common to all of these arguments against the state is that talk of the state having a positive role in green theory and practice belies an "environmental" rather than an "ecological" perspective (Bookchin, 1980, p.70, Dobson, 1990, p.13). That is, the idea of a "green" state is an oxymoron.

The strength of feeling generated by the place of the state within green politics can be most dramatically seen in the division within the green movement and green political parties between "fundi" and "realo" elements (Doherty, 1992). This division roughly coincides with the ecological/environmental dichotomy. Environmentalism is presented as a reformist strategy whose principal focus is on "greening" contemporary liberal democracies, rather than seeking the widespread social, economic and moral change necessary to create the "sustainable society", the aim of "ecologism" (Dobson, op cit., ch.3). For present purposes this distinction between environmentalism and ecologism will be assumed to centre mainly around opposing attitudes to the state, with ecologism being unequivocally anti state and environmentalism more agnostic.

Two anti state eco anarchist theories will be examined in this chapter: bioregionalism and social ecology. What I wish to argue is that the transformation rather than the abolition of the state is compatible with green values, and, given its wider concerns with issues of distributive justice and democratization, proffers a more accurate account of green aims than eco anarchism. Indeed, the argument is that as one moves along the continuum from bioregionalism to social ecology, the eco anarchist position "shades into" an understanding of green politics which sees its primary goal as the democratic transformation of the state and civil society. The argument, taken from theorists of radical democracy (Keane, 1988) as well as some eco socialists (Gorz, 1980), is that within the context of modern pluralist societies, democratization is intimately tied up with the maintenance of the division between the state and civil society. The radicalization of this traditionally "liberal" view of society, is what I take Eckersley's description of green politics being "post- rather than anti-liberal" (1992, p.30) to involve.

Eco anarchism

What O'Riordan has called the "anarchist solution" (1981, p.307) has been an enduring part of the green political spectrum. It is not difficult to see why anarchist social organization has appealed to greens. The basic anarchist claim is that left to themselves human beings will naturally cohere into "organic" communities regulated by principles of mutual aid and sociality (Sylvan, 1993, p.236). As such it coheres well with the naturalistic complexion of green politics. Social relations within stateless communities are such that the laws, procedures and institutions of centralized authority are unnecessary for governance. In short, the traditional defence of the state, that it alone can provide "public goods", particularly social order and environmental quality, is rejected.[1]

Bioregionalism begins from the argument that the resolution of ecological crisis calls for greater integration of human communities with their immediate environment, with natural rather than human political (read state) boundaries delimiting the appropriate human social unit or organizing its internal common affairs. This claim that the state is unnecessary for securing and enjoying public goods, I call the "weak" version of eco anarchism, to distinguish it from the stronger claims made by social ecology. While bioregionalism envisages an ecological stateless society, roughly along the lines of the traditional anarchist vision of a "commune of communes", it does not base its claims on a theoretically sophisticated assessment of the state. Social ecology, on the other hand, does.

For social ecologists, "The state consolidates and protects the family of hierarchies [class, gender, race, age, mind-body] becoming a hierarchy in its own right" (Social Ecology Network, 1992, p.8). Carter (1993) develops a comprehensive green anarchist theory which highlights many of the concerns of social ecology. Beginning from a view of the state as "an autonomous agent" (p.45), he claims the state cannot, nor ever could, be used to serve civil society (p.42). This discounting of the dominant "instrumentalist" view of the state, that is, seeing the state in functional terms as a "tool" of civil society, is the hallmark of what I want to call the "pure" anarchist position. On a pure anarchist reading the state has its own interests and agenda, and its sole raison d'etre is the systematic exploitation of society as a whole. Adding a feminist perspective, Bookchin holds that "The institutionalized apex of male civilization was the state" (1990, p.66). Social ecology thus advances what I call a "strong" version of eco anarchism. It goes beyond claiming that the state is unnecessary to stipulating that it is positively undesirable. From the strong or pure anarchist position, the state is not just part of the problem; it is the root of the problem. It is the Gordian knot, the severing of which is a necessary condition for the creation of an emancipated, ecologically rational society.

Central to both the strong and weak versions of eco anarchism then is the conviction that only the transcendence of state institutions, and their replacement with informal, community based social mechanisms, will guarantee the social conditions for the realization of green values. The anarchist argument is neatly summed up by Taylor: "If we want to do without the state or substantially reduce its role, we have to revive and rebuild communities" (1982, p.4). This distinction between strong and weak versions of eco anarchism is significant. The strong version of anarchism is problematic, particularly as regards the lack of empirical evidence to support its claims, while theoretically it has been criticized as dependent upon an ahistorical explanatory schema which confuses the concept of the state with particular conceptions. The pure anarchist position, the state as an intrinsically exploitative institution, is held by few greens. Indeed, as argued below, a close analysis of the social ecology position, specifically the "libertarian municipalism" of Bookchin, reveals that its aims can be accommodated by democratizing rather than dissolving the state.

Bioregionalism

Bioregionalists place a premium on the necessity of strong, affective senses of community and communal identity, and see the ecological problems we face as due in no small part to the demise of community. Although this decline in community is a common observation of communitarian critiques of, for want of a better word, "modernity" and the process of "modernization", bioregionalists locate one of the roots of its decline in the disengagement of people from the "commons". With no enduring link to the land, community becomes "rootless", hollowed out and functional, and individuals vulnerable to the anomie and alienation of "mass society".

186

For the nineteenth century sociologist Tönnies, the emergence of the modern social world was marked by a movement from "community" or gemeinschaft to a looser sense of "society" or gesellschaft (1957, p.33). Instead of individuals ruling themselves through traditional community structures, (external) state institutions became necessary to "order" an emerging "civil society" of individual competitive wills. This bioregional concern with recreating gemeinschaftlich is well expressed by Jones who declares that "If a future society based on the Gaian principles of interdependence, mutuality and interrelatedness is to be achieved, a re-cmergence of some form of gemeinschaft is essential" (1990, p.109).

Taylor (1982) spells out why anarchism demands communities of this type. For him stateless social order requires that the community display (a) strongly shared beliefs (pp.26-7), (b) relations between members that are direct and many sided (pp.27-8), and (c) social interaction characterized by reciprocity and mutual aid (pp.28-30). It is the quality of relations within such communities that allows the operation of non state coercive mechanisms to get people to do things they would not do voluntarily, yet are necessary for social order. A common misunderstanding of anarchism is that it implies social "chaos" and relies on a "myth of collective harmony" to coordinate that chaos. This myth is the belief that, once the state has been abolished, individuals will "automatically" or "naturally" be in harmony with one another. Neither claim is true. Realistic anarchists do not argue for the abolition of social coercion, but rather one particular form of it, namely the forms of institutionalized coercion employed by the state. Socialization, the internalization of communal norms and conventions together with forms of social sanction such as public ridicule and shaming, perform the necessary coercive functions within stateless communities (Taylor, ibid., pp.80-7). Thus the prevention of "free riding", and the provision of "collective goods", can be achieved without recourse to the state.

Recovering such communal principles, from a bioregional point of view, requires "reinhabitation", the conscious reintegration of human communities within their local bioregion. This is not just on the grounds that small scale communities living in close contact with "their' local environment are less ecologically destructive than large societies dependent on the biosphere as a whole. Reinhabitation, becoming "native" to a place, is held to be an identity constituting ecological condition. Who you are is a question of where you are, the types and quality of relations you find yourself in, both socially and ecologically.[2]

The basic bioregional vision is of a patch work of self sufficient, small scale, ecologically harmonious communities, organized according to their own normative standards (Sale, 1984a, p.233). Bioregionalists, as do many greens, discourage trade and stress the benefits of a high degree of communal autarky. One of the main reasons given for this is to encourage people to live within the limits set by their local environment rather than depending on the planet as a whole. Dasmann's distinction between "ecosystem people" and "biosphere people" captures the basic bioregional position. According to him, "Biosphere people draw their support, not from the resources of any one ecosystem, but from the biosphere...[They] can exert incredible pressure upon an ecosystem they wish to exploit...something that would be impossible or unthinkable for people who were dependent upon that particular ecosystem" (quoted in Sessions, 1992, p.121). Each ecosystem is unique, and thus demands a particular way of life. In this way the community's distinctiveness is intimately related to how it interacts with its ecosystem or bioregion. This interaction, and thus its identity, is codetermined by the ecological context, giving an ecological dimension to the communitarian worldview.

From the bioregional position, it is trade between disparate parts of the world that has created the illusion within those western nations who benefit most from planetary wide exchange, that escaping the finitude of their own particular ecosystem is synonymous with transcending "natural limits" as a whole. Economic growth beyond what would be possible if the economy was integrated with one ecosystem, its "natural hinterland", is largely based on trade. Surmounting the particular conditions of ecological scarcity has created the false impression that absolute ecological scarcity, entropic scarcity, is not a "given" but relative to human economic and technological conditions (Georgescu-Roegen, 1971, Lee, 1989). This is the ecological slant on the necessity for the imperialist expansion of western societies, and the ecological reason for present day neocolonial trade practices. Simply put, countries such as Britain, France and the United States could never have achieved the rates of material economic growth and affluence they did and continue to enjoy if they were dependent upon indigenous resources. The affluent lifestyle enjoyed in these countries is premised on a disproportionate consumption of world resources, and thus according to greens is impossible on a global scale. Bioregionalists encourage economic self sufficiency, which, by rendering the community dependent upon its local ecosystem, encourages a heightened sense of prudence, both moral and practical, in using it. In this way it seeks to save the whole (the global biosphere) by saving the parts (individual ecosystems), as well as fostering the independence and cultural uniqueness of bioregional communities.

There are positive aspects, from a green point of view, which can be drawn from bioregionalism: the emphasis on economic independence; ecosystemic sensitivity; and the environmental destruction and social exploitation attendant upon global trade. Despite these postive aspects there are problems with the bioregional vision. The first refers to issues arising from the lack of interaction between bioregions in the context of the distribution of resources across the face of the planet. Simply put, the autarky imperative, coupled with strict ecosystem dependence, implies that those living in resource poor ecosystems are condemned to their fate as there is no provision for the redistribution of resources between bioregions. The redistribution of resources across the planet, as envisaged by some as a core part of any green or environmentally informed theory of justice (Attfield and Wilkins (eds.), 1992), goes against the communitarianism which underwrites much of bioregionalism. Transfers, whether from trade or charity, may compromise the distinctiveness of bioregional communities, since their identity as a community is tied up with how they live within the possibilities of their ecosystem. Redistribution on any large scale, like trade, seems to be ruled out as homogenizing process, destructive of cultural difference and diversity. According to Berg and Dasmann, "Global Monoculture dictates English lawns in the desert, orange juice in Siberia and hamburgers in New Delhi. It overwhelms local cultures and "raises" them regardless of the effects on cultural coherency or capacities of local natural systems" (1981, p.25). Even "cultural" exchange, expressed in such practices as tourism, even if ecologically sensitive, is discouraged as being destructive of rootedness and communal distinctiveness (Mills, 1981, p.5). As such bioregionalism is arguably the most communitarian strand within contemporary communitarian theory.[3] An extreme interpretation would be that resource poor bioregions and communities have to simply survive and flourish as best they can on their own without any, or much, external exchange.

The bioregional vision of a world made up of self sufficient, inward looking, ecologically harmonious bioregions harks back to a premodern era before exploration, trade, and cultural exchanges brought people from different parts of the world together and gave tangible expression to the idea of the "human species". In place of the global village, with its communications networks, global political and economic institutions such as a world market, the World Bank and the United Nations, bioregionalism implies a refeudalization, or "Balkanization" of the world into ecologically defined political units (Sale, 1984b, p.171).

Other problems with bioregionalism can be grouped under two headings: those relating to internal relations; and external relations within and between bioregional communities. On the former, Dobson points out that there is no guarantee than bioregional communities will be democratic or just (op cit., p.122). Indeed, according to Sale "truly autonomous bioregions will likely go their own separate ways" (1984b, p.170). The reason for this is that bioregionalists place the communal right to self legislate as the highest social value. The affirmation of communal solidarity is prioritized over contingent values such as equality, fairness or democracy. Many of the problems with bioregionalism stem from the fact that neither "entry" nor "exit" is guaranteed, since any such guarantee would require some transbioregional authority. This communitarianism together with the assumption that the root of social and ecological problems is "bigness" (Sale, 1980, p.82), are the two principles around which bioregionalism is woven. However, small is not always beautiful, and small scale, although an important consideration, is not a panacea for all social and ecological ills.

An examination of how bioregionalism copes with internal differences brings into sharp relief its problematic place in green theory. Conflict within bioregional communities according Sale (1980), should not require recourse to formal principles of justice or political institutions external to the community. For example, the "natural" way to deal with disputes between an aggrieved minority and an implacable majority is for the community to divide, with the minority free to settle elsewhere. From the bioregional point of view "The commodious solution is not minority rights but minority settlements" (1980, p.480; emphasis added). This "fissioning" of communities when they get too large or develop tensions incompatible with communal consensus is, according to Taylor, "a normal part of the life of stateless societies" (1982, p.92, Sale, 1984b, p.170). But in a non-Lockean world, that is where there is no unsettled territory, this "solution" to conflicts within societies is simply unworkable. In a closed world, there is no "away" to which the displaced can go. This recourse to fissioning and relocating is surprising given the strong link made between communal and personal identity and the land. Perhaps the bioregional point is that people must make a choice between exile, thus compromising their identity, and putting up with the discomfort. In other words, they must rank community membership against other values and decide which is more important.

However, apart from these problems, one has only to look at the conflict within such divided societies as Northern Ireland, the former Yugoslavia and USSR to see that "minority settlements" do not constitute a realistic, never mind a "just", solution. Whether eco anarchists like it or not, the history of the nation state, and not necessarily the liberal constitutional

190

version either, provides ample evidence that it can protect the rights of minorities and individuals, as much as it can hinder them. The instrumental view of the state, contra the strong anarchist thesis, is not completely false.

This "justice as displacement" argument is premised on protecting the community's sense of identity and solidarity from those who argue for a different understanding of communal identity, shared goods, history or meanings: hence the distrust of appeals to justice as an entrenched system of individual rights and liberties which transcend local norms. Justice, as an ethical perspective that stands above communal conventions, is either incompatible with complete communal autonomy or is superfluous. Like other communitarians, bioregionalists, and perhaps some social ecologists, regard justice as a remedial virtue, useful for rectifying flaws in social life (Kymlicka, 1992, p.367): flaws that are the result of a decline in community.[4]

The danger here is not just the threat to individual liberty extreme forms of communitarianism may lead to, where the individual as an "organic part" of the wider collectivity can have her interests sacrificed for the benefit of the "common good". Rather, the conservative possibility inherent in aspects of the green position, illuminates the threat to plurality and social diversity as a precondition for democratic politics within contemporary conditions. This is because green communitarianism is usually presaged on communal self identity being constituted by a religious or quasi religious outlook. In bioregionalism this outlook is provided by deep ecology and the "reinhabitation" process, and has affinities with the "ecomonastic" strategy associated with Rudolf Bahro (Eckersley, op cit., pp.163-7). In both bioregionalism and ecomonasticism, the community is held together by a spiritualized view of the natural world in general and the local ecosystem in particular, and the community's relationship to it (Sale, 1984a).[5]

Given the post metaphysical nature of the modern world, it is extremely unlikely that a spiritualized view of the natural world, as opposed to a moral view, would succeed in "converting" western citizens to the green cause, though this is not to deny the efficacy of this approach in nonwestern societies. More damaging to green theory in terms of its "progressive" self understanding is that, historically, societies infused with a strong shared religious sense have typically been fertile breeding grounds for intolerance. A possible response is that the spiritualized worldview put forward by bioregionalists and deep ecologists is inclusionary, welcomes difference and otherness, a "unity in diversity" which comes mainly from its embracing of Eastern spiritual traditions. However, it remains to be seen if such a flexible, non dogmatic shared moral vision is sufficiently robust to furnish

191

the community with a strong shared identity in the sense required for stateless social mechanisms to work. Baldly put, there is every reason to believe that the tolerance proclaimed for green metaphysics may be undermined by the emphasis on "tribalism" (Sale, 1984b).

It is perhaps at the external level that the shortcomings of bioregionalism are most apparent. Given the on going impact of globalization on human societies, drawing them into an increasingly complex web of interrelations, the complete realization of the bioregional vision is impossible. Particularly when we look at the global nature of ecological problems such as ozone depletion and global warming, there is a need for more not less cooperation and interaction between societies. From a global ecological point of view, the fragmentation of the world as propounded by bioregionalism may exacerbate ecosystemic problems. The strategy of saving the whole by saving the parts only works if there is some degree of transcommunal cooperation and coordination. This is because when it comes to ecosystems, Commoner's first law of ecology holds; "everything is connected to everything else" (1971, p.29), saving the part involves knowing what is happening to other parts and to the whole. Although economic autarky may be possible, independence from the global ecosystem is not. Transbioregional problems imply that emphasizing individual ecosystem protection is a necessary but not a sufficient condition for saving the whole. Simply withdrawing from the global economy does not address how to solve existing commons problems although it may prevent them getting any worse. Given that trans societal coordination and communication is more important within such a context, decisions taken within bioregional communities are only meaningful within that context. Unfortunately, as Taylor admits, "the controls which can be effective within the small community cannot generally have a great impact on relations between people of different communities", (op cit., p.167). Added to this is Goodin's observation that decentralization gives each member of the community more control over that community's decisions. But the smaller the community, the less and less the community's decisions will ordinarily matter to the ultimate outcome. People are being given more and more power over less and less (1992, p.150).

In this sense Dasmann's conclusion that "the future belongs to...[ecosystem people]" overstates the bioregional case, to say the least (op cit., p.121). While not wishing to undermine the positive values expressed by the bioregional position, within the context of a globalized human species, institutionalized in the political, economic, and cultural interaction of sovereign nation states and various transnational communities, organizations and communication networks, it is not a perspective greens need take in toto. The positive aspects greens can take from it relate to the communitarian dimension of green politics, particularly the importance of "place" and its role as an identity forming condition, and decentralization and appropriate scale in fleshing out this dimension. The communitarian aspect of green political theory is also expressed by libertarian municipalism which we turn to next.

Bookchin's Libertarian Municipalism

Social ecology, although generally sympathetic to, and sharing much with, bioregionalism, offers a more coherent eco anarchist vision. Bookchin, the founder and leading theorist of social ecology, terms his vision of stateless social order "libertarian municipalism" (1986, pp.37-44, 1992a, pp.179-85, 1992b). This is defined as "a confederal society based on the co-ordination of municipalities in a bottom-up system of administration as distinguished from the top-down rule of the nation-state" (1992b, pp.84-5; emphasis added). It thus differs from bioregionalism in its concern with the issue of interaction between communities. The confederal nature of the arrangement makes it a voluntary association of autonomous communities with sovereignty retained at that level of the political arrangement. Yet, the relativism that typified bioregionalism is explicitly ruled out by Bookchin: "parochialism can...be checked not only by the compelling realities of economic interdependence but by the commitment of municipal minorities to defer to the majority wishes of participating communities" (1992b, p.97).

Libertarian municipalism as an eco anarchist theory can be argued to represent a novel form of anarchism. Limiting the scope of communities to simply go their own way marks a decisive break with traditional anarchist thought, which took the communal right to self governance as its principal and highest political norm. The distinction between libertarian municipalism and other forms of anarchism, including bioregionalism, has to do with their different understandings of community. Whereas for bioregionalists (Jones, 1990, Sale, 1980, 1984a), and "pure" anarchists (Taylor, 1982), community is understood as "gemeinschaft", Bookchin's political theory is presaged on the idea of a democratic community.

Another difference is that libertarian municipalism is urban rather than rural based (Bookchin, 1992a). Bookchin's understanding of community is thus less "organic" than traditional anarchist and bioregional views, which represents a development away from his earlier work which accorded normative significance to gemeinschaft, and praised the "authenticity" of "organic" forms of social life. One could question whether his current conceptualization of social ecology, libertarian municipalism, is anarchist in the traditional sense of the term rather than an attempt to spell out what a more democratized and decentralized society would look like with a continuing role for the state. For example, it is difficult to classify Bookchin's theory as anarchist when he states "Libertarian municipalism seeks to reclaim the public sphere for the exercise of authentic citizenship" (1992b, p.94). Such terms as "public sphere" and "citizenship" are more usually found within non anarchist theorizing about democratic theory and practice. The argument that "authentic citizenship" is only possible in his vision of political organization can be questioned as it is underwritten by an arbitrary, not to say simplistic, distinction between "statecraft" (representative, liberal democracy) and "politics" (authentic, direct democratic participation) (1992a, p.181).

His drift away from pure anarchism is further evidenced by his assertion that there is a "shared agreement by all [communities] to recognize civil liberties and maintain the ecological integrity of region" (1992b, pp.97-8). The contractual and legal-constitutional overtones of confederalism is more usually associated with liberal not anarchist discourse and practice. And his description of the confederal council as composed of elected representatives, with legitimate right to use coercion within a specified ecological territory, to ensure compliance with a shared agreement, could be taken as a traditional Weberian analysis of a state-like political entity, legitimated along standard, but beefed up, liberal practices (ibid., p.99). His assertion that the confederal councils, made up of deputies elected in direct democratic elections, are purely administrative with no mandated policy making powers, which is retained at lower levels (1992a, p.297, 1992b, p.97), is no more than that. Given the interconnectedness of communities, the existence of a binding confederal agreement relating to human rights and ecological imperatives, and the description of this social arrangement as a "Community of communities", one can imagine the confederal council taking a more proactive role than Bookchin assigns to it. Goodin's criticism made in reference to bioregionalism above applies a fortiori in this instance, since relations between communities go beyond trade or the maintenance of ecological integrity, but consist of transcommunal normative principles and practices.

A weak criticism of Bookchin's position would be that he has failed to demonstrate clearly the stateless nature of libertarian municipalism. Indeed, by conceiving the problematic of the state as a question of "degrees of statehood", rather than in monolithic terms of "the state", Bookchin's reformed eco anarchism is close to themes within

194

recent radical democratic theory, concerning the importance of plural and decentralized sites of political and social power independent from the state (Keane, 1988). This is particularly evident when Bookchin states that "the state can be less pronounced as a constellation of institutions at the municipal level, and more pronounced at the provincial or regional level, and most pronounced at the national level", (1992a, p.137), and seems to recommend city and local government level as appropriate sites for green activism which will not compromise its ends (ibid., pp.303-4).

A possible defence open to Bookchin's qualified anarchism is that it is a strategic step towards pure anarchism. According to Sylvan, "A committed anarchist can quite well also be committed, as an intermediate goal among others, to achieving more sympathico states. That, in turn, may involve political activity, conventional or unconventional", (op cit., p.241). This seems a fair assessment of municipalism as the latest conceptualization of social ecology, and its importance lies in the fact that social ecologists now share the goal of the democratic transformation of the state with non anarchist greens and democratic socialists, among others. It may be the aim toward which greens of different hues as well as other movements for social change can unite.

Part of this democratization project involves decentralization. The emphasis on small scale is a principle supported by almost all greens. It is usually taken as expressing the need for "appropriate scale" in decision making procedures and other areas such as production. It can be used to support a green argument for the state. According to Porritt "In terms of restoring power to the community nothing should be done at a higher level than can be done at a lower" (1984, p.166). This principle is compatible with state institutions because for some things, particularly international negotiation on global commons issues, it is the lowest level. The very term "municipal", with its strongly urban character, resonates and is compatible with the demand to strengthen local and regional tiers of government away from the centre. The principles of libertarian municipalism seem to accord with T.H. Green's assessment of those sceptical of the state. According to him, "The outcry against state interference is often raised by men whose real objection is not to state interference but to centralization, to the constant aggression of the central executive upon local authorities" (1974, p.217).

Devolving power to municipal levels, yet maintaining the legitimate right for the "confederal council" to intervene in municipal affairs, makes the latter a state like institution based on an ecological social contract, "the shared agreement" (1992b, p.98). In the manner of a decentralized and democratized state, it circumscribes communal rights to

195

complete self legislation, since upholding the ecological compact depends on such circumspection. Political power is shared rather than completely devolved to local levels, which, as Bookchin presents them, do not seem possessed of sufficiently strong senses of common identity in the manner of gemeinschaftlich, which could underwrite complete communal self governance in the manner of pure anarchism or bioregionalism. What I want to suggest then is that the libertarian municipal agenda, the content of which most greens would accept, such as participatory democratic structures, local empowerment, social justice, and human rights, is more consistent with a political project aimed at democratizing the state and civil society.

Conclusion: eco anarchism: from constitutive to regulative ideal[6]

Why has the eco anarchist vision of a federated community of small scale, face to face communities living in harmony with the environment been such an enduring feature of green political theory? To answer this question one cannot divorce green political theory and green political practice. It is a praxis oriented political ideology, one concerned with convincing and persuading others of the merits of its case in the various domains of social life, the "public sphere" and media, culture, and at the formal political (state) level. Why then has eco anarchism been so influential within green politics ?

In the first place, envisioning a future "sustainable society" vividly captured in shorthand form the basic principles of green politics: inter alia, ecological and social harmony; decentralization; simple living; quality of life; community; and democracy. Early formulations of green theory simply assumed that an ecological transformation of society required anarchism updated for the age of natural limits. Secondly, the dominance of the eco anarchist vision has to do with the dichotomous style adopted by green theorists and commentators. The most influential instance of this is O'Riordan's fourfold typology of the institutional choices open to green politics: (1) new global order; (2) authoritarian commune; (3) centralized authoritarianism; or (4) the anarchist solution (1981, pp.404-7). In reality the choice comes down to the anarchist solution or the rest, given that it was the only one which guaranteed the values and principles noted above. As it developed therefore, it was again assumed that the "sustainable society" was "anarchistic" (Dobson, op cit., pp.83-4).

In a way this utopia building, supported by positive references to preliterate aboriginal societies which "proved" the case that stateless societies were more ecologically

196

sustainable than ones with state structures, was prompted by the need for greens to adopt a strongly critical edge in their analysis of contemporary industrial societies. As Goodwin points out, "the process of imagining an ideal community, which necessarily rests on the negation of the non-ideal aspects of existing societies, gives utopian theory a certain distance from reality which makes it a sharper critical tool than much orthodox political theory" (1991, p.537). One could say that its initial reaction to the contemporary social world was so antithetical, so radical in questioning almost every aspect, that utopianism was the only form of theorizing which could contain and convey the green message.

This anticipatory utopian form of political critique is directly related to the evolution of the green movement, and its roots are manifold. Firstly, the practical requirement as a "new social movement" to maintain its distinct identity, to prevent existing ideologies from coopting its ideas, presented a good case for accentuating the radical, the utopian. Secondly, in common with other new social movements, greens seemed to be particularly obsessed with questions of self identity, to demonstrate to themselves as much to anyone else their "newness". A typical example is Porritt's declaration that "For an ecologist, the debate between the protagonists of capitalism and communism is about as uplifting as the dialogue between Tweedledum and Tweedledee" (1986, p.44). This statement is doubly noteworthy as it not only lumps these two alternatives together as simply different versions of the super ideology of "industrialism", but also compounds this by equating socialism with communism. In contrast to socialism/communism and capitalism, greens were "post" or "anti industrial", which cast them as the vanguard of the future society. Hence their slogan "neither left not right, but in front". Thirdly, added to these internal dynamics was the simple fact that as new social actors, they had little or no access to the policy making process, and therefore did not need to outline programmes, budgets or detailed policies. Broad brush strokes rather than attention to the fine print characterized early green discourse. The overriding imperative was to distance themselves theoretically from the reality surrounding them (and in the case of Bahro's eco-monasticism, to literally turn one's back on the existing social order). This formative experience, like all formative experiences, still exerts a strong influence on green politics. Fourthly, this concern with outlining their diametric opposition to the status quo, was underpinned by the naive belief that the green case was so obvious and compelling that all that was needed was to simply express it (Dobson, op cit., p.23). Finally, following the "doom and gloom" that typified the post *Limits to Growth* ecology movement, there was clearly a need felt by greens to outline an image of a better future if they were to persuade people to the green cause. Thus, like a skilled preacher, the early green movement had threatened apocalypse if its warnings were not heeded: now it also promised the eco anarchist, liberated society, if people changed their ways.[7]

A cursory review of green literature will quickly highlight the extent to which green theorists and commentators are obsessed with presenting the green case in an either/or format. Almost ubiquitous is the habit of drawing lists distinguishing the "green" from the putatively "non green". Examples include Porritt's two 29 item lists differentiating "The politics of industrialism" from "The politics of ecology" (op cit., pp.216-7), O'Riordan's technocentric/ecocentric dichotomy (1981, ch.1), Dobson's distinction between "ecologism" and "environmentalism" (op cit., p.13), to Capra's "paradigm shift" from "The Newton World-Machine" to "The New Physics" (1982, part II). Surprise that this dualistic methodology is so widespread within a political theory that is supposed to be holistic is only surpassed by the fact that it persists to frame its concerns. It is from this dualistic methodology, coupled with the utopian-critical demands of the early green movement, that eco anarchism became the dominant political theory of greens. Three steps can be identified in this process:

1.   The concern with mapping what a "sustainable society" would look like to highlight the unsustainable nature of existing society led to,

2.   A focus on mapping "the sustainable society", that is describing, often in great detail, a generally agreed picture/blueprint of that society,

3.   Finally the assumption of the sustainable society as "anarchistic", to rule out eco authoritarian dystopias, and to act as the benchmark against which "greenness" could be judged.

It is the particular historical development of green theory (both internal debates, and between it and other theories such as socialism and liberalism), and green political practice (fundi/realos), that largely account for the predominance of the eco anarchist solution. These factors produce a marked tendency within green theory to work backwards, as it were, from utopia to theory, with practical engagement in the political realities surrounding it reduced to publicly articulating the utopian-theoretical synthesis. Although there is nothing wrong with outlining a vision of a better society, indeed this prescriptive dimension is the mark of any ideology worth its salt, this tendency unfortunately resulted in the

198

description of the future society becoming a substitute for underwriting green theory. The task of specifying and spelling out green principles has only recently been undertaken (cf. Dobson and Lucardie (eds.), 1993). These principles were assumed to be self evident, the need to justify them lessened by the fact that they could be "read off" from the future eco anarchist society. An analysis of green principles may reveal that there is no reason to believe that a society consistent with them will necessarily be "anarchistic". It is perhaps more than coincidence that the common "reading off" social principles from nature often occurs together with "reading" them off from the eco anarchist society, one reinforces the other (cf. Dobson, op cit., pp.24-5, Sale, 1980, pp.329-35, Bookchin, 1991, pp.75-86).

What this paper hopes to have shown is that eco anarchism, as a constitutive ideal of green politics, is neither necessary in that the values greens espouse can be institutionalized in non anarchistic ways. Nor can it be desirable that, as the relationship between green political theory and eco anarchism stand, the latter could be said to act as a fetter on the former, unnecessarily precluding, as this essay has intimated, a positive engagement between green theory and the state. In critically assessing the eco anarchist vision, and seeing the importance of the state in terms of realizing green principles and values, what we are perhaps witnessing is the transition of green political ideology into green political theory. It is perhaps not completely contingent that this recent development within green theory occurs at a time when greens are serious contenders for political power, when the minds of greens are turning from ideals to principles and from principles to practice. This is not to say that eco anarchism is to be rejected from the green political canon: the integration of its insights within the current context of green politics moving from negative criticism to positive proposals calls for it to become a regulative rather than a constitutive ideal for green politics. That is, informing and guiding, but not determining its goals.

Notes

* The author wishes to thank Piers Stephens and Diane Sinclair for their comments and editorial assistance on earlier drafts of this paper.

1.   For some "deep" greens, eco anarchism is understood as a return to the "natural" social set up for humans. This type of argument is prominent within "neo primitivist" shades of green theory, and others who stress the significance of stateless societies as the social order within which the human species evolved (Shepard,

1992). Here the ascription of positive moral value upon such societies is presaged on a rather simplistic identification of the "natural" with the "good".

2. Another overlap between bioregionalism and social ecology is the idea that the historical evolution of the state as a social institution is linked to the transition from a Neolithic hunter gatherer way of life, to a sedentary, agricultural society. With the rise of agriculture came the possibility of large social surpluses which in turn sparked off the expansion of formal institutions of social coercion and appropriation (cf. Bookchin, 1990).

3. A further objection to bioregionalism relates to dangers inherent in presaging communal identity and membership on the strong affinity between the community and "the land". This is the "brown" or fascistic interpretation of "reinhabitation", which may be construed as a green version of "blood and soil" nationalism. This volkisch tendency within certain strands of green theory is noted by Vincent (1993, p.266) and Coates (1993).

4. A witty example of green communitarianism is the following limerick by Boulding (in O'Riordan, 1981, p.33),

    Economists argue that all the world lacks is
    A   suitable   system   of   effluent   taxes
    They forget that if people pollute with impunity
    This must be a symptom of lack of community.

5. This emphasis on the self and the constituent conditions for identity formation make bioregionalism close to the deep ecology position in this regard. In affirming a contextual understanding of self and community, with the ultimate context being the environment, bioregionalism states, albeit in a rather extreme form, one of the central goals of green politics. This is to conceive of the self as "self-in-society-in-environment", a recovery of a naturalistic account of the self. On this point green politics has much in common with the communitarian critique of the "individualism" and false understandings of the self within liberal theory.

6. The distinction between constitutive and regulative ideals is taken from Kant who held that "A principle is "regulative" when it merely guides our thinking by

indicating the goal towards which investigation should be directed...it is "constitutive" when it makes definite assertions regarding the existence and nature of the objectively real", (1957, p.211).

7. The Marxian critique of utopian socialism is an obvious analogy here, and indeed many Marxists have criticized green politics, or some conceptions of it, as a modern day version of utopian critical theory (Pepper, 1993). The basic Marxist critique of green theory is that it lacks a political economy, a theory of transition to the "sustainable society", a vital political need that utopia building does not fulfil. Engels' rejection of the utopian socialists expresses one of the basic Marxist problems with green politics, "To all these [utopian socialists] socialism is the expression of absolute truth, reason, and justice, and has only to be discovered to conquer all the world by virtue of its own power" (1978, p.693).

References

Attfield, R. and Wilkins, B. (eds.) (1992), *International Justice and the Third World: Studies in the Philosophy of Development*, Routledge, London.

Berg, P. and Dasmann, R. (1981), "Reinhabiting California", in Berg, P. (ed.), *Reinhabiting a Separate Country: A Bioregional Anthology of Northern California*, Planet Drum Foundation, San Francisco.

Bookchin, M. (1980), *Towards an Ecological Society*, Black Rose Books, Montreal.

Bookchin, M. (1990), *Pathways to a Green Future*, South End Press, Boston.

Bookchin, M. (1992a), *Urbanization and the Decline of Citizenship*, Black Rose Books, Montreal.

Bookchin, M. (1992b), "Libertarian Municipalism: An Overview", *Society and Nature*, vol. 1, no. 1.

Capra, F. (1982), *The Turning Point*, Fontana, London.

Carter, A. (1993), "Towards a Green Political Theory", in Dobson, A. and Lucardie, P. (eds.), op cit.

Coates, I. (1993), "A Cuckoo in the Nest: the National Front and Green Ideology", in Holder, J. et al. (eds.), *Perspectives on the Environment: Interdisciplinary Research in Action*, Avebury, Aldershot.

Commoner, B. (1971), *The Closing Circle: Nature, Man and Technology*, Bantam Books, New York.

Dobson, A. (1990), *Green Political Thought*, Unwin Hyman, London.

Dobson, A. and Lucardie, P. (eds.) (1993), *The Politics of Nature: Explorations in Green Political Theory*, Routledge, London.

Doherty, B. (1992), "The Fundi-Realo Controversy: An Analysis of Four European Green Parties", *Environmental Politics*, vol. 1, no. 1, Spring.

Eckerlsey, R. (1992), *Environmentalism and Political Theory: Toward an Ecocentric Approach*, University of London Press, London.

Engels, F. (1892), "Socialism: Utopian and Scientific", reprinted in Tucker, R. (ed.) (1978), *The Marx-Engels Reader*, 2nd ed., WW Norton, New York.

Georgescu-Roegen, N. (1971), *The Entropy Law and the Economic Process*, Harvard University Press, Harvard.

Gorz, A. (1980), *Ecology as Politics*, Verso, London.

Goodin, R. (1992), *Green Political Theory*, Routledge, London.

Goodwin, B. (1991), "Utopianism", in Miller, D. et al (eds.), *The Blackwell Encyclopaedia of Political Thought*, Blackwell, Oxford.

Green, T.H. (1881), "Lecture on Liberal Legislation and Freedom of Contract", reprinted in Diggs, B. J. (ed.) (1974), *The State, Justice and the Common Good*, Scott, Foresman & Co., Glenview, Illinois.

Heilbroner, R. (1980), *An Inquiry into the Human Prospect*, WW Norton, New York.

Jones, A. (1990), "Social Symbiosis: a Gaian Critique of Contemporary Social Theory", *The Ecologist*, vol. 20, no. 3, May/June.

Keane, J. (1988), *Democracy and Civil Society*, Verso, London.

Kant, I. (1957), "Critique of Pure Reason", in Greene, T. (ed.), *Kant: Selections*, Charles Scribner's Sons, New York.

Kymlicka, W. (1993), "Community", in Goodin, R. and Pettit, P. (eds.), *A Companion to Contemporary Political Philosophy*, Blackwell, Oxford.

Lee, K. (1989), *Social Philosophy and Ecological Scarcity*, Routledge, London.

Mills, S. (1981), "Planetary Passions: A reverent anarchy", *CoEvolution Quarterly*, vol. 32, no. 4.

O'Riordan, T. (1981), *Environmentalism*, 2nd. ed., Pluto, London.

Ophuls, W. (1977), *Ecology and the Politics of Scarcity*, W.H. Freeman, San Francisco.

Pepper, D. (1993), *Eco-Socialism: From Deep Ecology to Social Justice*, Routledge, London.

Porritt, J. (1984), *Seeing Green: The Politics of Ecology Explained*, Basil Blackwell, Cambridge.

Sale, K. (1980), *Human Scale*, Secker & Warberg, London.

Sale, K. (1984a), "Mother of All: An Introduction to Bioregionalism", in Kumar, K. (ed.), *The Schumacher Lectures* Vol. 2, Blond & Briggs, London.

Sale, K. (1984b), "Bioregionalism: A New Way to Treat the Land", *The Ecologist*, vol. 14, no. 4, July.

Sessions, G. (1992), "Ecocentrism, Wilderness, and Global Ecosystem Protection", in Oelschlaeger, M. (ed.), *The Wilderness Condition: Essays on Environment and Civilization*, Island Press, Washington and Covelo, Ca.

Shepard, P. (1992), "Post-Historic Primitivism" in Oelschlaeger, M. (ed.), op cit. *Social Ecology Network*, Paper no. 1.

Taylor, M. (1982), *Community, Anarchy and Liberty*, Cambridge University Press, Cambridge.

2 Democratic theory and environmental protection

James Meadowcroft

As its title suggests, this paper will be concerned with two themes: "theories of democracy" and "environmental problems". By "theories of democracy" I mean basic politico theoretic understandings of what democracy is, how it works, and why it should be of value to us. By "environmental problems" I mean the tangled knot of issues brought to the fore in the second half of the twentieth century by the ever increasing scale of the human impact on the non human natural world.

Although it is possible to trace the roots of modern environmentalism back more than a century,[1] it is only over the past few decades that environmental issues have emerged as a major locus of political controversy.[2] Hardly surprising then, that until recently,[3] democratic theorists have had very little to say explicitly about the way in which environmental dilemmas are framed, and can be resolved, within democratic political systems. But while issues we now consider "environmental" did not figure prominently among the problems which the understandings of democracy elaborated by earlier generations of theorists were intended to address, it is possible for us to bring the two together; that is to say, we can confront the established theories with the new problems to see how they fare.

With its focus on *theories* of democracy, this enquiry can be considered a tangential approach to the more pressing practical issue of the extent to which real world democratic polities are equipped to deal with environment related problems. In so far as the theoretical perspectives I will explore capture something of how representative democracies actually operate, difficulties displayed by the theories may suggest something about the difficulties of real world polities.

I shall begin by sketching out the general contours of three theories of democracy. Then I shall introduce a number of typical "environmental problems", considering how they would be understood, and how they could be resolved, within the horizons established by each of the various theories. Essentially I shall argue that none of the accounts commonly offered by theorists implies a representative democracy that would be especially adept at resolving environmental problems.

Theories of democracy

Over time, political theorists have developed a bewildering variety of approaches to democracy. For the purposes of this paper, I will concentrate on a limited number of influential twentieth century perspectives. Rather than focusing on the particular views of individual theorists, I will present three "ideal type" descriptions.[4]

The democratic practices and institutions of which the various theories offer an account are assumed to be those typical of modern "representative democracies": political systems based on the rule of law, where governments are determined by periodic multi party elections based on universal suffrage or a close approximation thereof, which maintain extensive individual liberties, and rights of free association and a free press.

In each case, I will say something about the essential vision of democracy embodied in the theory, about the perspectives it explicitly rejects, and about its perception of individual electors and political parties. It should be noted that each of these theories has an explanatory and a normative dimension, simultaneously offering an account of how representative democracy actually works, and a justification for why this is the way it should work.

Popular government

Democracy is understood to be rule by the people, a system in which the community as a whole takes responsibility for determining the laws under which its citizens will live, and orders the priorities of government according to the shared understanding of the common good. Freedom of speech and association, and competitive elections, allow a free airing of alternative viewpoints and expression of the perspectives of various groups and interests. In democratic discussion individuals and collectivities must justify their claims by an appeal to what is right and good for the community. Through the process of debate and exchange a common understanding of how best to promote the common welfare can be built, and governments can apply policies generally appreciated to be for the benefit of the community as whole. Of course, specific interests always seek to tilt the balance of public decision making in their favour, and it is even possible for the people as a whole to be misled by demagogues or to become fixated on false ideals; but the free press, the right for minorities and opposition currents to organize, and the periodic election of government means that these dangers can be minimized.

In this vision of democracy, the function of the politician is to grasp the authentic needs and desires of the community, which may be partially obscured by a mass of competing and contradictory particular interests, and to propose policies congruent with this popular will. Political parties contribute to formulating the alternatives that face the community. They focus debate on real choices. Individual electors are naturally most concerned with issues which touch them directly; but the excitement of electoral contests, based upon party competition, stimulates a more general interest in public affairs, and forces consideration of the policies and personnel best suited to the time. Democracy is not here simply the rule of the majority, or of the largest faction; nor is it simply a mechanism through which competing interests are accommodated; rather it is process through which an interest common to the community can be constituted and hold sway over public institutions.

Democratic elitism

Democracy is understood as a system which allows for the selection and renewal of political elites through a process of electoral competition. Elections provide a process for choosing a governing team, and for dismissing it at periodic intervals should it prove incompetent, excessively corrupt, or become divorced from the concerns of the electors. The work of governance and political decision taking should remain the preserve of a political elite. The mass of the electorate lacks the knowledge to make reasoned judgements about complex political affairs. Nor do they have the inclination to devote the time and attention necessary to acquire such knowledge. The ordinary voter is prone to consider issues emotionally, to be swayed by non rational arguments, and to be both inconsistent and overly rigid. The actual work of politics is best conducted by specialists; by professional politicians who acquire skill and experience, who can appreciate the complexity of domestic and international problems, who learn what can and cannot be accomplished, and understand how to accommodate opponents and make compromises. Parties assume an important functional role within such a system. They channel the political expression of diverse groups and interests, formulate policy options and mould opinion, constrain and discipline political actors, focus political competition, and present the electors with a choice of alternative governing groups.

Thus the virtue of modern democracy is not that it is "rule by the people", but that it provides: a competitive arena within which political personnel can gain experience and contend for power; a framework for relatively open debate on issues of public concern; avenues through which diverse groups and interests can strive to influence policy choices; a mechanism to constrain the influence of the state bureaucracy and to prevent the permanent capture of governmental organs by particular vested interests; and a means to effect the peaceable transition of power from one governing team to another. It is simply not possible for the people "to rule" in any real sense, and theories of democracy which suggest that they can do so are profoundly mistaken. Of course, the myth of "popular rule" which such theories propagate can enhance the legitimacy of democratic regimes, but should this be taken seriously by politicians or electors could lead to disastrous consequences. An excess of participatory zeal would place strains on the system, making the accommodation of conflicting interests more difficult and reducing the political specialists' freedom of manoeuvre. From this perspective a relatively small political elite represents the ingredients of an ideal democracy. The members of this elite, divided along party lines, may compete among themselves for power but accept the fundamental workings of the system. This elite will be surrounded by a wider group of party activists who mobilize support and provide candidates for promotion to leadership groups. The whole system is embedded in a more or less quiescent population which merely bestirs itself to vote periodically.

Pluralist democracy

Democracy is understood as a set of mechanisms which facilitate the organization, expression, and conciliation of group interests. Policy outcomes depend upon a complex interaction of diverse, differentially organized, partially overlapping and fluid groups. As the relative significance of interest aggregations change over time, so the main directions of public policy will evolve. Groups may be organized by an appeal to a variety of bases including locality, ethnicity, religion, socio economic class, sphere of economic activity, sex, sexual orientation, special needs, leisure pursuits and so on. Groups and interest aggregations vary greatly in size and stability, financial and organizational strength, and according to the degree to which their objectives are central to the self identity of their constituencies. Powerful economic interests may have a strong influence over policy formation; but all groups can wield some influence, and no group is so powerful that a coalition of other forces cannot bring it to book.

208

Precisely because there are so many splintered and partially overlapping groups, democracy is not prone to domination by a single fixed "majority"; the working majorities of the political system are no more than temporary coalitions of diverse minorities that come together at a particular time to achieve limited objectives. No single interest could dominate for long and no coalition of interests is ultimately stable.

Political parties aggregate interests into coalitions broad enough to win office and serve as channels through which groups can influence policy decisions. The individual voter is important as a locus of multiple identities and as an adjudicator over the extent to which different party coalitions have successfully assimilated the perspectives of the groups to which he or she is attached. The function of the politician is to represent, and then reconcile, diverse and often contradictory interests. On one hand politicians are pressured to defend specific interests, on the other excessive identification with one group may hinder the coalition building necessary for election.

The apparatus of the government itself is seen to be fractured into different alignments and rife with bureaucratic wrangling. Societal interests establish close links with relevant departments such as farmers with the ministry of agriculture, arms manufacturers with the ministry of defence, and so on. Government itself becomes involved in internal processes of interest accommodation and compromise.

Thus, democracy is not "rule of the people" because there is no one "people", but rather a mosaic of overlapping groupings. Nor is it rule by an elite, because politicians and parties are ultimately reactive, reflecting the strains of conflicting group interests.

Environmental problems

Problems which can be broadly described as environmental now constitute a central focus of political controversy. In part this is a question of redefinition; we now regard as "environmental" issues which previously might have been understood as problems of worker "health and safety" or efficient "resource management".[5] But there is no doubt that the conceptual constitution of a class of environmental problems has a material foundation in the vast increase in the scale of human impact on the non human natural world which has occurred during the second half of the twentieth century. This a phenomenon results partly from an intensification of industrial techniques which have been with us for some time such as dam building and fossil fuel combustion and partly from the development of new and more exotic technologies as well as from the dramatic increase in human numbers.

Environmental problems often appear particularly intractable because they seem to require us to make choices, sometimes irrevocable, between incommensurate "goods": the scenic beauty of a forest and the jobs of lumber workers; the survival of an endangered species; or wealth to be extracted through mining. More importantly, *environmental problems are typically experienced as external constraints which frustrate established expectations* and which require an adjustment of existing social practices; this adjustment entails costs; and it provokes struggle over how these costs are to be distributed.

So how would democracies, functioning according to the descriptions provided by our three theoretical perspectives, handle environmental conundrums? In each case the general mechanism by which a newly perceived or emergent problem could be tackled is relatively straightforward. Under *popular government*, the process begins when one section of the people identifies an abuse or problem; perhaps because of specialist knowledge, or because they experience its effects more directly. Active elements then initiate agitation for reform, seeking to rally the citizen body to their cause. A few of the more imaginative political leaders then come to take up the issue. As pressure for reform mounts, vested interests threatened by proposed solutions begin to mobilize, trying to bring into play the community's instinctive resistance to change. Eventually, one of the major parties is converted to reform; and finally the bulk of the people is galvanized. In the face of a frantic rear guard action by opponents, government acts. Ironically measures are often taken by the party which initially resisted reform. Within a few years the new order appears so natural that no one can really understand how things could ever have been done differently. Under *democratic elitism*, political leaders are first to grasp the political ramifications of a problem brought to their attention: perhaps by their contact with other functional elites such as leaders of business or the scientific community; by their knowledge of international trends; or by their ongoing supervision of domestic social processes. Elite teams will consider possible solutions and integrate alternatives into their broader plans for government. Because the issue will form but one part of the basis on which voters will choose a ruling team, the political elite will have considerable latitude as to the particular solutions adopted and the timing of change. Under *pluralist democracy*, an issue reaches the public agenda when taken up by one or more groups whose interests are directly affected. As the severity of the problem grows, a larger range of interests is implicated, and the intensity of the public debate will grow. Politicians will be pressured by all concerned. Initially reform proposals may be traded off in bargaining over more pressing matters. If the issue is sufficiently serious to mobilize a broad spectrum of interests in favour of action, government will institute reform, although perhaps in a watered down form which takes some account of the interests threatened by change.

To assess the kind of a response these mechanisms would actually generate, let us consider how polities functioning along these different lines might respond to four rather typical environmental problems.[6]

Case one: acute urban smog

Let us assume the urban centres of our test polities to be regularly and increasingly effected by serious smog resulting from high particulate and gaseous emissions from a variety of industrial and residential sources. More than an inconvenience, low air quality is contributing to a high incidence of respiratory disorders and excess deaths, particularly among the older population.

Perhaps not surprisingly, each of these democratic systems could be expected to respond relatively promptly to this challenge and to adopt some form of pollutant emission controls. Reformers in the *popular government* polity would have a specific and highly visible problem on which to fix their attention. The identification of a "good" common to all is relatively straightforward: every urban inhabitant breathes the polluted air and suffers the accompanying discomfort and added health risks. Vulnerable groups are dispersed across the population, although they may be particularly concentrated among the economically disadvantaged who have generally poorer health, less opportunity to escape the cities or adjust work schedules. Since all citizens have a direct experience of the problem, to galvanize the community for action reformers need only establish that the problem is rectifiable and that the cost of so doing is not prohibitive.

In the *elite democracy* polity, the issue touches the whole population directly. The effects on public health can be established easily. The problem is readily perceptible since anyone can look out of a window and conclude that something is wrong and the effectiveness of remedial action can be rapidly assessed which would encourage political elites to place the issue high on their list of priorities.

In the *pluralist democratic state*, the key lies in the balance of interests: here, interest in rectifying the problem would be widespread, growing in intensity in proportion to the severity of the smog. Interests opposed to clean air legislation could include industrial concerns facing costly adjustments to their production methods and others who disapproved on more general grounds to the intrusion and red tape which would accompany a new regulatory regime. Resistance by threatened firms would be insufficient to block action to

211

remedy an acute problem, but it might well win government subsidies to help defray the costs of pollution control technology.

Probable reaction is, of course, closely dependent upon the structure of the hypothesized problem. Should a) the anticipated costs of rectificatory action be raised, or b) the transparency of the problem and of the remedial measures be reduced, then the probability of government action is diminished: this across all three of the democratic models under consideration. For example, consider this variant: once again the problem is smog, but here confined to a small town with one large industrial employer. Fear of damaging the interests of the company with which the economic prosperity of the community is linked may indefinitely postpone the introduction of regulative controls: local government may not resolve to act, while national government remains preoccupied with other priorities. In terms of our model polities: at the local level the *common good* seems best served by leaving things as they are, the municipal political *elite* opts for the status quo, and the interests favouring change remain weak as compared with the coalition of company, union and small business *interests* opposing it; at the national level the *common good* is defined in terms of more general objectives, the local issue does not rank highly in the electoral preoccupations of the national *elite* teams, and no wider *interest* coalitions are brought to bear on the issue.[7] Or, let us return to a nationwide phenomenon but vary the nature of the air contaminant: say we are dealing not with smog, but with lead from gasoline combustion. The fact that here the presence of the pollutant is not readily visible, and that the damage done to human health and to the environment is more difficult to establish, could be expected to delay the introduction of regulatory measures.

Case two: natural gas conservation strategy

Let us assume substantial quantities of natural gas have been discovered within the territorial limits of our model democratic states. Reserves are sufficient to last forty years if the rate of exploitation of the resource is determined entirely by market variables. This is predicated on a significant expansion of gas usage as new gas power stations are brought on stream to replace older, more expensive, as well as environmentally pernicious, coal fired plants. However, some economists and environmentalists urge government to intervene to slow the pace of extraction to double the anticipated lifetime of the reserves.

The chances the "conservation" option will receive serious consideration in the *popular democratic* polity appear thin. The issues involved are complex and technical, requiring a careful measurement of gains and losses, and there is considerable uncertainty. The conservation policy option rests on the fact that gas is a clean, flexible fuel in limited abundance; by artificially restricting supply in the present, more of the resource is set aside for future consumption when its scarcity value will be enhanced. The argument for "immediate use" emphasizes current benefits to industry and consumers in the form of lower electricity prices and the immediate reduction in sulphur dioxide and carbon dioxide emissions. Besides, who can predict circumstances forty years hence? By then new gas supplies may have come on stream or other energy sources become available. Furthermore, the accelerated economic activity provided by cheaper energy today will generate compounding growth, building a richer society better able to cope with the challenges of the future. The determination of the common good is therefore far from straightforward. Even if a clear intellectual case can be made for conservation, it is highly unlikely that the people as a whole could be galvanized to act. We are not dealing with a problem with acute or immediate effects since there is no glaring abuse to focus opinion; rather it is a matter of abstract argument and counter-factuals, not the sort of issue around which the popular will can readily coalesce.

Things would work out similarly for the *pluralist democratic* polity. Strongest support for a regulated release of the gas reserves would presumably come from coal producers who would face declining sales as the switch went ahead. Considerable public sympathy might arise for the fate of those traditionally employed in this industry. Environmentalists might well be split over the issue, some preferring immediate emissions reductions, others preferring resource conservation. On the other side, gas producers would favour a rapid expansion of markets to maximize the return on their investment; utilities would favour larger supplies of a cheaper fuel; and groups representing industrial and domestic electricity users would support the switch to gas. Furthermore, in an economic environment where supply side decisions were usually left to producer groups, the adoption of an interventionist regime to fix maximum production levels would be highly unlikely.

On the face of it, the *elite polity* would appear to have a slightly greater chance of giving the "conservation" strategy a fair hearing. Precisely because the issue is somewhat abstract, it would be unlikely to be a major determinant in how voters cast their ballots when reviewing their choice of governors. Despite this relative freedom of manoeuvre, however, it is unlikely that the outcome would be enormously different. Political elites are in intimate contact with other elites, especially the business elite on whom they are to some

213

extent dependent as a decline in investor confidence could seriously damage the economic climate, reducing the re-election prospects of that political team. Because of the immediate economic benefit of lower electricity costs and reluctance to see increases in government regulation, it is unlikely that the business elite would support gas production ceilings. Furthermore, the elite government knows that voters will judge it on current economic performance, rather than projected benefits to future generations; its tendency will be to bring into play all the factors which could improve economic performance and maximize economic growth before the end of its mandate. Should the economy turn sour, to have backed the "conservation" strategy would be to have left oneself open to a charge of having artificially aggravated the recession; but by the time any problems of the "immediate use" strategy show up, this generation of political leaders will be long gone.

Case three: estuary development

Let us suppose a proposal to stimulate economic activity in an economically depressed coastal region by constructing an eight mile barrage across the mouth of a river estuary. The barrage will suppress tidal fluctuations and create a large body of water which can serve as the focal point for a major resort/marina development project. An added benefit of the plan is that it would eliminate spring tide flooding of farm land higher up the river.

In the *pluralist* democratic polity the complex array of interests touched by this scheme would soon be revealed. Financial and engineering concerns sponsoring the project would offer vigorous support. Backing would also come from a wide coalition of local businesses, such as hoteliers, restaurants and shop owners, who would gain from the major influx of vacationers into the area. The large number of jobs to be created during the five year construction phase, as well as the long term openings to be secured in the tourist industry, would probably contribute to substantial local support. Property values in the area immediately adjoining the estuary could be expected to rise, creating a windfall for land owners. Finally, farmers in the higher reaches of the river would be pleased with the reduction in flood risk. Opponents of the project might include: local residents who feared that the barrage and the influx of tourists would disturb traditional patterns of life; local fishermen whose livelihood was threatened since the absence of tidal flows would obliterate shellfish populations in the river mouth; and conservation groups concerned with habitat destruction, particularly the elimination of feeding zones for migrating birds.

The final outcome of this clash of interests is impossible to predict. On the face of it, supporters of the development project appear much the stronger as a large array of actors would stand to make significant economic gains. But note that while some large companies would be involved, support for the project is above all concentrated locally. Other regions of the country and firms not directly involved would gain little from the project and would loose little if it did not go ahead. A well organized campaign by environmental groups might well mobilize sufficient pressure at a higher jurisdictional level of national or regional government to slow down or halt the project. After all, for the general public no sacrifice is involved if the project fails to proceed, while there is the potential gain of preserving the coastal region and wildlife habitats.

Of course, other factors could effect this balance. The depressed coastal region might be seen as a financial drain on the rest of the country and the project would then appear as a way of keeping down general taxation, although on a project of this modest scale potential savings to the central budget would probably be negligible. The wider public might empathize with the economic plight of the coastal dwellers and so favour development, but perhaps these concerns could be alleviated by encouraging a less environmentally intrusive form of local development. People in other regions and companies not involved in the project might dislike central intervention to stop the project in case it established a precedent which threatened projects in which they did have a more direct interest.

The ultimate outcome would depend upon the reaction of these and many other factors, including the political and legal balance of power between local and regional governments. If the project did proceed, it might include compensatory mechanisms for some of the damaged interests such as cash or alternative jobs for the fishermen, preservation of a modest alternative site for the birds and so on.

In the *elite democratic* polity, much would depend on the relationship between the concerns of the national and local elites. If local and national elites concurred, the project would move rapidly ahead. If they disagreed, then the mechanisms for settling local/centre conflict would become important. This is unlikely to be an issue on which a national election would turn, but if the political position of the national elite team were weak it might have to consider carefully whether the loss of conservation votes spread across the country, assuming it let the project go ahead, could harm it as much as the loss of seats in the disadvantaged coastal constituencies, assuming the project was blocked in the face of strong local support.

The *popular democracy* might experience this division as a disjunction between the local and national perceptions of the

215

common good. Theoretically, the good as determined by and for the whole people is more significant than that of any fraction of the whole. However, there are inevitably many partial goods which sections of the people share; and sometimes it is for the good of the community as a whole to secure goods for its more disadvantaged members. It is not at all clear that in this case, however, the rather defuse general desire for "conservation", provided others assume the costs, felt by the bulk of the community should trump the intensity with which the local community desires jobs and economic well being. Thus once again it seems that the popular democracy would have difficulty identifying to true bearing of the common good.

In each of the three polities, therefore, the tangle of considerations becomes so complex that the outcome cannot be reliably anticipated. Of course, if the example were varied somewhat so that the net perceived national advantage in the development scheme was substantially increased then, other things being equal, the probability of the development proceeding would be enhanced. This might be the case if the issue was oil drilling rather than resort building; or that the tourists were to be foreigners bringing in badly needed foreign exchange, for example.

Case four: transgenetic manipulation

Let us assume that significant advances in human ability to manipulate genetic material have laid the basis for the commercial development of transgenetic organisms and that the leaders of our test polities must consider how to formulate an appropriate regulatory regime.

The problem for the *popular democratic* polity lies in the difficulty of conducting an informed public debate. Advocates of the new technology would emphasise the potential benefits to humankind such as the possibilities of high yield and disease resistant crops, more productive livestock, new medical treatments, and the replacement of polluting industrial processes by environmentally benign ones. Critics would stress dangers and unforeseen consequences; they would ask about the ethical, legal and social implications of manipulating human and non human genetic material, about possible military applications and so on. But to the vast bulk of the population the question would remain abstract and inaccessible. The policy which government adopted towards transgenetic manipulation would, at this stage, have no direct impact on their lives. Above all, the great technical complexity of the issue and the rapidity with which knowledge in the area is changing would make it difficult for the average citizen to make an informed judgement. Thus, in this case, public opinion would be unable to coalesce around a clear conception of the common good. The leadership of the popular democracy would therefore have little choice but to follow the natural inclination of their colleagues in the elite democracy and turn to the scientific experts most knowledgeable about the emergent technology. On the basis of their advice, a policy of prudent encouragement would likely be adopted. Work should continue since to hesitate would be simply to hand leadership in the field to others, but under relatively strict regulative procedures; procedures managed, necessarily, by those most familiar with this domain.

Similar courses of action would probably be adopted by the *elite* and *pluralist* democracies, although the confidence in specialized elites and the preference for settling matters away from the public gaze, typical of the former, might lead to greater reliance on "self regulation" by the industry; while the more open collision between supporters and opponents characteristic of the latter might lead public officials to adopt a slightly more interventionist regulatory regime.

Note that in each polity a general debate on whether society really needs or wants this technology is forestalled, or more accurately it becomes irrelevant. Decisions about the direction of research and product development are basically left to scientific and commercial interests. Products enter the market, after passing the appropriate regulatory hurdles, one by one; many will go almost unnoticed; the odd one will spark a broader public debate; but over time the technical foundation of the society will shift and new political, social, and ethical practices will gradually become established. Note also that in each polity those who have the greatest professional and financial interest in commercial applications of the technology are most likely to play the central role in the regulatory regime.

Conclusion

What conclusions can be made on the basis of this brief thought experiment? First, with respect to the three theories of democracy which we have been discussing, it does seem that the policy outcome adopted to deal with a specific environmental problem may be effected by the kind of ideal type democracy in which the decision is made. The government in an elite democracy has more room for manoeuvre having a greater capacity to take decisions independent of the immediate currents of public opinion or balance of interests. However, the relatively closed nature of elite politics may slow the identification of emergent problems. Popular democracy has difficulty dealing with issues on which there is no clear public consensus, but once it is resolved to act, it may act speedily and decisively. Pluralist democracy is both responsive to a wider range of concerns and more fluid and erratic; it can work out accommodation based on a complex array of interests, but a relatively small change in the pattern of interests can lead to a rapid erosion of a pre-established consensus. In many cases such differences may bear upon the possible triumph of a particular environmental policy outcome.

On the other hand, it is evident that the observed differences are rather small. More importantly, they do not bear consistently in any particular direction: that is to say, the peculiarities of these democratic variants are not such that the operation of one system more consistently generates satisfactory outcomes to environmental challenges. In other words, the elite's freedom of manoeuvre, the popular democracy's carefully constructed resolve, and the pluralist democracy's complex interest coalitions can, in different contexts, tell equally for or against the successful management of environmental problems.[8]

Should this outcome surprise us? In one sense probably not: after all, these are rival understandings of the operation of a similar set of institutional practices such as competitive multi party elections, free press and individual rights. Furthermore, these political mechanisms are situated within the context of a socio economic structure and a prevailing set of ideological assumptions which closely constrain the policy options open to decision makers. On the other hand, considering the intensity with which partisans have fought over these visions of democracy, it is interesting that when judged according to the straightforward utilitarian criterion of differential outputs in handling environmental issues, there seems little to choose between them.

References

1. See for example: Pepper, D. (1984), *The Roots of Modern Environmentalism*, London; Bramwell, A (1989), *Ecology in the Twentieth Century*, New Haven and London; and Taylor, B. (1992), *Environmental Political Thought in America*, Kansas.

2. Lester, J. (1989), *Environmental Politics and Policy*, Durham, NC; Robinson, M. (1992), *The Greening of British Party Politics*, Manchester; Weale, A. (1992), *The New Politics of Pollution*, Manchester.

3. Recent contributions include Paehlke, R. (1988), "Democracy, Bureaucracy and Environmentalism", *Environmental Ethics* 10; Dryzek, J. (1992), "Ecology and Discursive Democracy", *Capitalism, Nature, Socialism* 9; Goodin, R. (1992), *Green Political Theory*, Cambridge;

Saward, M. (1993), "Green Democracy", in Dobson, A. and Lacardie, P. (eds) (1993), *The Politics of Nature*, London.

4. For background to these three constructions see Pennock, J. R. (1979), *Democratic Political Theory*, Princeton; Held, D. (1987), *Models of Democracy*, Cambridge; Arblaster, A. (1994), *Democracy*, (Milton Keynes); and, Birch, A. H. (1993), *The Concepts and Theories of Modern Democracy*, London and New York.

5. See for example: Yearley, S. (1991), *The Green Case*, London; and Simmons, I. G. (1993), *Interpreting Nature*, London and New York.

6. For the purposes of this analysis I assume that we are dealing with otherwise similar, contemporary, medium sized, prosperous, industrial states, with largely privately owned and market mediated economies.

7. Note that there is a close structural parallel between this "one industry town" scenario, as experienced on the local level, and the situation confronting a national polity, operating in an international regime of free trade, which declines to introduce more elaborate environmental controls for fear of eroding its position in international markets.

8. Throughout this discussion I have assumed that these test polities work more or less as their advocates claim. One could imagine distopian versions of these three models: where the general will was always determined in an atmosphere of prejudice and was hopelessly fickle and erratic; where elite teams were equally incompetent, or where the obvious unsuitability of one team gave the other a semi permanent monopoly on power; where one interest monopolized decision making, or the fractiousness of groups produced endless stalemate. However, it is not clear that any one of these "deformed" regimes would consistently generate a worse environmental record.

3 Subsidiarity and global warming policy: An excuse for inaction in the European Community?

Ute Collier

Introduction

The months of protracted negotiations over the Treaty on European Union, commonly known as the "Maastricht" Treaty, after the Danes' initial rejection of the Treaty in May 1992 had at least one positive aspect. It actually created some awareness amongst the average citizen of the current issues surrounding European integration thanks to the intensive media coverage surrounding the Treaty negotiations. Suddenly, "Maastricht" was in (nearly) everybody's mouth and with it the then latest Eurobuzzword: "subsidiarity". However, few people actually seemed to understand what the European Union (EU) bureaucrats meant by the term. In fact, bureaucrats and politicians themselves seemed equally unsure and, at one stage, Commission president Jacques Delors offered a price of 100,000 Ecu for the best one page definition of subsidiarity.

In simple terms, subsidiarity means that action should be taken at the appropriate level. A number of countries had been campaigning for some time against the, in their view, excessive centralization of decision making at EU level. With the revision of the Treaty of Rome through the Single European Act in 1987, the EU had been officially given responsibility for a number of policy areas ranging from agriculture, via economic and social cohesion, to research and technology development. Some member states felt that within these spheres of responsibility, directives were being agreed which might as well be left to national or even local legislation. There was a feeling that the EU was excessively bureaucratic and the popular view was of millions of EU bureaucrats wasting huge amounts of time and money on issues like common rules on the depth of tractor tyres.[1]

Subsidiarity was made one of the guiding principles of the EU through its inclusion in the Treaty on European Union but it has actually applied to environmental policy since the 1987

Single European Act. Now there was increasing concern that "subsidiarity" could be interpreted in ways that could be negative in environmental terms. While some member states have been very progressive in their environmental policy and have in fact implemented stricter action than that required by EU legislation, others have been dragging their feet, in particular the southern member states. "Subsidiarity" could be used as an excuse for inaction by such countries.

The aim of this chapter is to examine the environmental implications of the subsidiarity principle with the EU's climate change strategy as a case study. The paper begins with a discussion of the subsidiarity principle and its ramifications for EU environment policy. It then outlines the main features of the EU's climate change strategy and examines how the subsidiarity principle has influenced its development. The next section contains a short review of the climate change response strategies of some of the member states. Conclusions are then drawn about the prospects for climate change abatement in the EU.

Subsidiarity and EU environment policy

It is apparent that policy decisions have to be taken at various levels of government. There is little value in central government deciding issues relating to, say, waste collection which is obviously a local issue while at the same time foreign policy is obviously a national issue. However, there are few issues where the division of responsibility is so clear cut. Hence, in individual EU countries, as elsewhere, the areas of competence of the various levels of government vary significantly. Some countries, such as the UK and France, are much more centralized than others, especially federal countries like Germany and Belgium. Even within countries, the division of responsibilities can be open to debate and change. The UK government for example has over the last decade or so consistently removed powers from local authorities.

Not surprisingly, therefore, there are different views on the exact role of the European Union. While some, like the then Commission president Jacques Delors, pursue visions of a federal Europe, others, like the UK government, are concerned about their loss of sovereignty. These different viewpoints are nothing new but seem to have become amplified in recent years, in particular since the Single European Act (SEA), which amended the Treaty of Rome in 1987, broadened the EU's areas of responsibility. The SEA contained the first references to the idea of subsidiarity. Article 130r, which gave a legal basis to EU environment policy, mentioned that:

The Community shall take action relating to the environment to the extent to which the objectives referred to in paragraph 1 can be attained better at Community level than at the level of the individual member states. Without prejudice to certain measures of a Community nature, the Member States shall finance and implement the other measures (ECSC-EEC-EAEC, 1987).

As the impetus for greater European integration grew in the late 1980s, Eurosceptics increased their calls for a clear division of responsibilities. Hence, according to Verhoeve, Bennett and Wilkinson (1992), during the early stages of negotiations for the Treaty on European Union, some member states and members of the European Parliament argued that the principle of subsidiarity should be extended to all EU activities and be defined in some detail, so as to clarify the distribution of competences and to enable the principle to be enforced by the Court of Justice. However, in the event, no precise definition was included. Article 3b of the Treaty states:

In areas which do not fall within its exclusive competence, the Community shall take action, in accordance with the principle of subsidiarity, only if and in so far as the objectives of the proposed action cannot be sufficiently achieved by the Member States and can therefore, by reason of the scale or effects of the proposed action, be better achieved by the Community. Any action by the Community shall not go beyond what is necessary to achieve the objectives of this Treaty (Council of the EC, CEC, 1992).

As mentioned above, the driving force for the implementation of the subsidiarity principle came mainly from a concern about excessive bureaucracy and centralization related to a whole variety of EU activities. However, there have also been suspicions that some member states have pushed subsidiarity because they are concerned about a possible loss of sovereignty and have become indignant about the Union's growing role in many policy areas, including environment policy. Verhoeve at al (1992) argue that member states might invoke the subsidiarity principle and more frequently insist on a convincing justification for new proposals, particularly as the new Treaty allows qualified majority voting (QMV) on most environmental measures. This means that individual member states have less scope for blocking proposals in the traditional way of vetoing them. However, at the same time, the imprecise definition of the principle would make any challenge in the Court of Justice unlikely to succeed, in particular as there is generally a good case why environmental action needs to be taken at a supranational level.[2]   Verhoeve et al. are thus likely to be correct in their assumption that in practice the implications of the subsidiarity principle are less a matter of action or no action but mainly a question of the nature and extent of Union action.

In general, subsidiarity does not necessarily have to have negative environmental implications.   Centralization of decision making has been one area which environmentalists have seen as a particular problem. The Green movement has long campaigned for greater citizen responsibility.   The potential value of decentralization has also been stressed within the context of sustainable development.   The World Commission on Environment and Development (WCED) for example, which was instrumental in bringing the concept of sustainable development onto the political agenda, felt that there needed to be greater public participation in decisions that affect the environment. According to them:

> This is best secured by decentralizing the management of resources upon which local communities depend, and giving these communities an effective say over the use of resources (WCED, 1987).

This aim appears to be met by the EU which, in the Treaty on European Union, states that decisions should be taken as closely as possible to the citizen. Sustainable development has been made the central principle of the EU's environment policy, in particular since the publication of the Fifth Environmental Action Programme (EAP) which is entitled "Towards sustainability".   In it, reference is made to "subsidiarity":

> The principle of subsidiarity will play an important part in ensuring that the objectives, targets and actions are given full effect by appropriate national, regional and local efforts and initiatives (EC, 1992a).

The Fifth EAP gives a new dimension to subsidiarity, namely the concept of "shared responsibility":

This concept involves not so much a choice of action at one level at the exclusion of others but, rather, a mixing of actors and instruments at the appropriate levels, without any calling into question of the division of competences between the Community, the Member States, regional and local authorities (EC, 1992a).

Hence, it looks that, at least in theory, the principle of subsidiarity could actually be compatible with the aims of better environmental protection and sustainable development.[3] Its realization is likely to be along the lines of the EU setting a framework for action, with more detailed policies being drawn up at national, regional and local levels, depending on where action is appropriate. While in principle this sounds a sensible solution, considering the deplorable environmental record of certain member states, this may result in a lack of action. There has been ample evidence over recent years that while some countries, like Denmark, Germany or the Netherlands, have been very proactive in their approach to environmental problems and have often initiated EU directives, other member states, generally the southern member states and the UK, have been dragging their feet. One wonders whether the subsidiarity principle is necessarily invoked with environmental considerations in mind and whether it will actually result in the appropriate level of action. The principle could be used as an excuse by national governments to maintain their sovereignty or, even worse, to scale down their responses to environmental problems.

The aim of the next section is to examine a relatively new area for EU environment policy, namely the climate change issue, and ascertain to what extent the subsidiarity principle is having an impact on policy developments in this area.

The EU's climate change strategy

Obviously, climate change is a global environmental problem and the climate convention agreed at the UNCED in 1992 is a first step towards global action on the problem. Global agreements have to be implemented through action at the national and local level but action at EU level can also be justified. To start with, the twelve EU countries now constitute one of the most powerful economic blocks in the world and can thus attempt to exert pressure on other countries to act on environmental issues. Furthermore, it is useful to coordinate action between countries and to exchange information. Finally, certain measures such as appliance standards or energy taxes need harmonization as a prerequisite for the proper functioning of the Internal Market.

As a first measure, the member states agreed on a joint target for emission reductions. The aim for the EU as a whole was to stabilize CO_2 emission levels by the year 2000. However, some member states are committed to actual emission reductions while others, mainly the cohesion countries, expect an increase in emissions. The approach is one of "burden sharing". When it came to drawing up a CO_2 strategy, disagreements on a number of issues emerged, with different viewpoints from the various Directorate-Generals involved, as well as from the member states. Originally, a whole range of specific measures was envisaged but consecutive drafts have seen a significant scaling down of the proposals. The Commission eventually published a first Communication on the issue in October 1991, followed by a second Communication in June 1992.

The Communication (EC, 1992d) was accompanied by proposals for four specific measures as follows:

- a framework directive on energy efficiency (SAVE);

- a decision concerning the specific actions for greater penetration of renewable energy resources (ALTENER);

- a directive on a combined carbon/energy tax;

- a decision concerning a mechanism for monitoring of Community CO_2 emissions and other greenhouse gases.

The 1991 Communication assumed the need to reduce emissions by 11 per cent from the 1990 level to achieve stabilisation (EC, 1991). The 1992 Communication revised this figure upwards to 12 per cent due to an accelerating growth in emissions in 1991, which denotes that there may be some further problems ahead with higher than expected increases. The different measures and programmes were expected to contribute different proportions of the required reductions as can be seen in Table 1:

Table 1: Projected emission reductions from the different components of the Community strategy

Proposed measures reduction for stabilization	Expected CO_2
Carbon/energy tax	6.5%
SAVE	3.0%
THERMIE	1.5%
ALTENER	1.0%
Total	12.0%

Source: EC, 1992b

While, initially, it looked as if the EU would take strong action on climate change, the following sections will show that the prospects are now less certain. This is to some extent due to the subsidiarity doctrine being invoked. The two measures that have been most affected are the carbon/energy tax and the SAVE programme.

The Carbon/energy tax

A main focus in the development of the Commission's strategy on CO_2 has been on the possibility of introducing a tax in order to internalize some of the external costs of energy. Pressure for an EU level tax came from the fact that three member states, Denmark, Germany and the Netherlands, were threatening to introduce carbon taxes unilaterally, thus infringing the Commission's attempt to harmonize taxes for the proper functioning of the single market. It was decided that there should be no sole CO_2 levy as this would have favoured nuclear power which some member states opposed.[4]

First proposals for a tax were put forward in a communication to the Council in late September 1991. The proposed tax was a so called hybrid carbon/energy tax amounting to $10/barrel by the year 2000, starting with $3/barrel as of 1 January 1993 and increasing by $1 annually. Further details were put forward in a communication in June 1992. The first casualty was the proposed starting date, which by then was no longer mentioned.

According to the 1992 proposals, the tax is to be based half on CO_2 emissions (expressed in tonnes) and half on the calorific value of the fuel (expressed in gigajoules). Energy from renewables (except hydropower plants above 10 MW) is to be exempted from the tax. It was clear that the tax

227

proposals would attract opposition from various industrial groupings. Intensive lobbying against the tax by industrial groups has taken place, accompanied by threats of moving industrial production outside the EU. As a result a number of concessions have been made which, if applied, would substantially weaken the effect of the tax. Firstly, member states may be authorized to grant tax reductions up to 75 per cent to firms whose energy costs amount to at least 8 per cent of the value added of its products and whose competitiveness would be threatened by the tax (EC, 1992c). Member states would also be allowed to grant temporary total exemptions to firms that have embarked on "substantial efforts to save energy or to reduce CO_2 emissions" (EC, 1992c).[5] This vague statement is liable to lax interpretation and the exemptions have seriously compromised the effectiveness of the tax as they mean that the largest consumers of energy in the EU would pay the lowest rates of tax, thus giving them little additional incentive to invest in energy efficiency.

Entry into force of the carbon/energy tax was made subject to the adoption of similar measures by the EU's main trading partners, namely Japan and the United States. The proposals specify that the tax should be revenue neutral, that is it should not result in any net increase in statutory contributions and charges. Environmentalists have not been satisfied with this arrangement and have suggested that the revenue should be earmarked for use for environmental purposes, such as energy efficiency incentives. The Commission's argument, according to officials interviewed by the author, is that the revenue would be much too large to be used in this way. In any case, the proposals do include the option of using some of the revenue for providing tax incentives (EC, 1992c). One problem here is that in the UK, for example, the Treasury will not allow revenue to be earmarked for special purposes and is unlikely to amend this position. Essentially, the Commission cannot force member states to use the tax for particular purposes, so, once again, the effectiveness of the taxation would depend on the goodwill of the member states.

Subsidiarity has been used as one of the justifications for opposing the tax. The UK has been at the forefront of the opposition and has been arguing that it would be more appropriate to develop a tax at the national level. Since its decision to impose Value Added Tax on domestic fuel in March 1993, the UK government claims that it has already instituted a form of carbon tax. However, by following the UK government's general attitude to EU matters, it is clear that a main reason for objecting to the tax is not a concern about subsidiarity but a general reluctance to surrender decision making powers to the EU, especially on an important matter such as taxation. At the same time, other member states, in particular the cohesion countries, are concerned about the effect of the tax on their competitive position while France wants to see a sole carbon tax in order to protect its nuclear industry. Germany and the Netherlands, who are staunch supporters of the tax, have responded by blocking progress towards the EU's ratification of the UN climate convention.[6] These disagreements, under the influence of the subsidiarity principle, have thus created a logjam and there is much uncertainty as to whether agreement on the tax can ever be reached.

The SAVE programme

Improvements in energy efficiency have to be a quintessential part of any CO_2 reduction strategy. Initially, SAVE was proposed to enable the EU to meet its 1995 energy objective of improving efficiency by at least 20 per cent. First proposals for SAVE were published under separate cover in November 1990 and became subsequently integrated into the CO_2 strategy. The proposals of November 1990 envisaged a variety of measures under SAVE, as illustrated in Table 2. The measures marked with an asterisk are still contained in the final directive published in 1993 but action will be left to the member states (see below). Only those marked with @ have seen specific EU level action.

Table 2: Proposed and actual measures under SAVE

- » minimum insulation standards for new buildings (*)
- » energy certification of buildings (*)
- » billing of heating costs based on actual consumption (*)
- » promotion of third party financing (* but public sector investments only)
- » inspection of boilers (*)
- » energy audits for businesses (* but only those with high energy consumption)
- » efficiency standards for water heaters (@, directive adopted 1992)
- » energy labelling (@, directive adopted 1992)
- » pilot studies on least cost planning (@, currently underway)
- » performance levels and possibly efficiency standards for refrigerators and freezers
- » minimum performance requirements for cars
- » a general speed limit of 120 km/h
- » removal of obstacles to the development of combined heat and power
- » energy taxes
- » road pricing
- » coordination of member states' activities

Source: EC, 1990; EC, 1993

As can be seen, some of the measures proposed initially have simply been dropped. The idea of a speed limit, for example, was almost immediately abandoned, due to total opposition to this measure from Germany. Progress has been made with some measures, such as the energy labelling directive (EC, 1992e). Most significantly, SAVE has been turned into a so called framework directive, which means that the EU only sets the general principles for action, on which member states then have to base their programmes of measures. This has been a direct result of member states invoking the principle of subsidiarity.

The final directive thus states that the member states shall draw up and implement programmes in the six areas retained in the final directive but the member states essentially have a free hand in designing and implementing programmes. The directive declares that:

Programmes can include laws, regulations, economic and administrative instruments, information, education and voluntary agreements whose impact can be objectively assessed. (EC, 1993)

The rest of the directive contains similarly vague requirements. The Commission proposals for SAVE from 1992 still were somewhat more specific, for example requiring the certification of public sector buildings at a rate of at least 5 per cent of the existing stock per year (EC, 1992d). The final directive just talks generally of energy certification of buildings. At the same time, the proposals have been watered down even further. The inspection of heating installations is now restricted to those above 15 kW and energy audits are only required for industrial undertakings with high energy consumption rather than businesses in general.

The member states have to report to the Commission every two years on the results of the measures taken and the effectiveness of SAVE will only emerge in a few years' time. An examination of the policies in the member states can provide a guide to the likelihood of action at that level. The next section will thus give a quick review of the activities of individual member states.

Climate change strategies in the member states

As Table 3 shows, a number of member states have actually set themselves individual CO_2 targets which go beyond the general EU target.

Table 3: CO_2 targets of individual member states

Country	Target	Base Year	Commitment Year
Belgium	5% Reduction	1990	2000
Denmark	20% Reduction	1988	2005
France	Stabilization	1990	2000
Germany	25-30% Reduction	1987	2005
Greece	EU agreement		
Ireland	EU agreement		
Italy	Stabilization	1988	2000
	20% Reduction	1988	2005
Luxembourg	Stabilization	1990	2000
	20% Reduction	1990	2005
Netherlands	Stabilization	1989/90	1995
	3-5% Reduction	1989/90	2000
Portugal	EU agreement		
Spain	Limitation to 25% Growth	1990	2000
United Kingdom	Stabilization	1990	2000

Source: IEA. 1992

These targets are one indication that most member states are prepared to act on the climate change issue. However, as the examination of the EU level policy indicates, it is crucial to look beyond the official target to the actual implementation of policy measures. The author has carried out extensive research on the climate change strategies of three member states, namely Germany, the Netherlands and the UK. This has revealed varying levels of commitment to action. It is beyond the scope of this chapter to enter into any detail on the various response strategies, so only the main points are mentioned here (for more detail see Collier, 1994).

Germany, while having set itself the toughest target for CO_2 reductions, has been rather slow to introduce policy measures. Unification has been a heavy drain on the country's finances and has resulted in political difficulties for the current government so that climate change has been pushed from the top of the agenda. There is for example a lack of support for energy efficiency at the federal level.

However, the federal nature of government in Germany means that lower levels of government have responsibility for certain decisions in the energy and environment area. A number of states and local authorities have been much more active than the federal government. This has generally been in the context of the development of so called energy concepts, which aim to integrate planning of energy demand and supply, with a particular focus on energy efficiency and renewable energy. These concepts have not been exclusively designed with CO_2 reduction in mind, but they have nevertheless a major beneficial effect.

The Netherlands is another member state which has been proactive on environmental issues. While its target is not as ambitious as Germany's, it is actually well in the process of implementing it. An important part of the Dutch strategy has been the agreement of voluntary covenants with a number of industrial sectors including the energy sector. In the latter case, the companies have been allowed to add a small levy to consumers' bills in order to finance their programmes which include various energy efficiency measures and investment in combined heat and power and renewables (see Collier, 1993). Furthermore, there are a number of government subsidies available to promote renewables and combined heat and power. The Dutch government's approach essentially meets the EU's ideas about shared responsibility with government, industry and non governmental organizations involved in the formulation and implementation of policy.

As already mentioned, the UK has been one of the countries most obstructive to reaching agreement on the EU CO_2 strategy. Its particular objection has been to the carbon/energy tax as it feels that decisions on taxes should not be taken at EU level. Instead, it has recently decided to impose Value Added Tax on domestic fuels which the government claims is part of its CO_2 strategy although the opposition and environmental groups are very sceptical, believing that the main reason for its imposition was to help to reduce the large budget deficit the country faces. In terms of energy efficiency, currently most support measures are aimed at businesses, thus missing out a large proportion of energy consumers. Basically, the UK government's main preoccupation in recent years has been with the privatization of the electricity sector which has resulted in companies in which decisions are dominated by profit considerations. Under the new system, some support has been made available for renewable energies, although the most supported energy source is nuclear power which receives most of the receipts of a 10 per cent surcharge on electricity prices. Due to the central structure of decision making in the UK, subsidiarity

233

almost entirely depends on effective measures being taken at central government level and some doubt has to be raised about the commitment of the government to environmental issues.

As far as the other member states are concerned, a similarly mixed picture can be found as to the implementation of CO_2 targets. A good review is provided by IEA (1992). Some countries like Denmark are actively implementing CO_2 limitation strategies while others like Greece do very little and rely on the fact that, as it has been accepted that their economies still need to grow, they will be allowed to increase CO_2 emissions. As other work by the author (see Collier, 1994) has demonstrated, the type and effectiveness of policy response to the climate change issue depends on a whole variety of political, regulatory and institutional factors which are not easy to determine. Most member states are actually implementing measures to reduce CO_2 emissions, so that the failure to agree on measures at EU level does not imply total inaction. Nevertheless, the EU member states are currently unlikely to achieve the stabilization target for 2000. When the member states submitted their plans for CO_2 reductions to the Commission, they cumulatively missed the target by 4 per cent. One problem is certainly that southern member states actually predict large growth rates for CO_2 emissions.[7]

Conclusions: appropriate levels of action or an excuse for inaction?

The above discussion has shown that the implementation of the subsidiarity principle has influenced the development of the EU's strategy to deal with climate change. In terms of the SAVE programme, subsidiarity has meant that the directive now only provides a framework for action rather than more specific measures such as appliance standards. How such a framework directive is to be enforced is uncertain. Considering there have been problems with enforcing the implementation of "proper" directives, the general vagueness of framework directives can only increase the enforcement problem. As far as the carbon/energy tax is concerned, there was still no sign of an agreement by mid 1994, nearly three years after the initial proposals.

Most followers of the political wrangles surrounding policy developments at EU level will be sceptical if member states claim to invoke subsidiarity for environmental reasons, suspecting that subsidiarity might just be used as an excuse for inactivity. It is of course not easy to prove or disprove such claims. The fact is that with the signature of the climate convention, each EU member state has committed itself to drawing up national programmes to mitigate climate change. Hence, all member states are now committed to some kind of action on climate change. However, some would be unlikely to adopt the type of measures contained in the EU strategy, in particular the carbon/energy tax. So, while the subsidiarity principle does not necessarily serve as an excuse for inaction, it may nevertheless be used as an excuse when the real issue at stake is a fundamental objection to certain measures such as taxes. In the case of the carbon/energy tax, the point is not necessarily an objection to this specific measure but to the general principle of the EU making decisions on taxes which some governments see as a threat to their sovereignty.

As a final point, the question remains whether in the areas under discussion, action at the national, regional or local level might indeed be more appropriate. As mentioned, energy efficiency, especially on the end use side, has to be a crucial component of any climate change strategy. A successful end use efficiency programme will have to influence the consumption behaviour of individual energy consumers and Hennicke, Johnson, Kohler and Seifried (1987) have found that energy efficiency programmes can be organized most effectively at the local level. Furthermore, desirable energy technologies in environmental terms, such as renewables and combined heat and power are often small scale and are thus more suited to local energy companies rather than centralized monopolistic companies.

There is thus a case for allowing flexibility in decisions on the best means of promoting energy efficiency and renewables and leaving scope for decision making at lower levels of government. Hence, the framework directive approach may have its advantages. The only drawback is that as certain member states seem to be somewhat reluctant to commit themselves to environmental action, the EU Commission would have less scope to coerce the member states to action than with more specific directives. Hence, the overall conclusion is that subsidiarity and environmental protection are not necessarily incompatible objectives. Indeed, the idea of "shared responsibility" as advocated in the Fifth EAP is a step in the right direction. However, the problem is that, as so often, in reality such principles often get abused by member states, either as an excuse for inaction on environmental issues or as a means to protect their power and sovereignty.

Notes

1. In reality, the Commission of the EU actually only has a modest staff compared to many of the member states' bureaucracies and in 1992, only 2.3 per cent of the overall EU budget went on administration.

2.   This relates to arguments such as pollution knows no boundaries or the need to harmonize product standards.

3.   Sustainable development is itself a somewhat ambiguous concept which implies different things to different people. However, it is beyond the scope of this chapter to enter the discussion about the exact meaning of "sustainability".

4.   In particular Italy, where the then Environment commissioner Ripa di Meana originated.

5.   Additionally, the Council, acting unanimously on a proposal from the Commission, can suspend the application of the tax in "exceptional cases in order to take account of the special situations in member states".

6.   as reported in *ENDS Report* 222, July 1993, pp.40-42.

7.   Greece and Spain expect emissions to rise by at least 25 per cent, Portugal between 29 and 39 per cent (as reported in *Environment Digest* 68, February 1993, p.10).

References

Collier, U. (1993), "Global warming and energy policy in the Netherlands: towards sustainable development?", *Dutch Crossing*, Winter 1993, pp.69-91.

Collier, U. (1994), *Global warming and electricity supply: towards the integration of energy and environment policies in the European Community?*, D.Phil. thesis, University of Sussex.

Council of the EC, CEC (1992), *Treaty on European Union*, Office for Official Publication of the European Communities, Luxembourg.

EC (1990), "Proposal for a Council Directive concerning the promotion of energy efficiency in the Community", *COM* (90) 365 final.

EC (1991), "A community strategy to limit carbon dioxide emissions and to improve energy efficiency", Communication from the Commission to the Council *SEC* (91) 1744 final.

EC (1992a), "Towards sustainability", *COM* (92) 23 final.

EC (1992b), "A community strategy to limit carbon dioxide emissions and to improve energy efficieny, Communication from the Commission", *COM* (92) 246 final.

EC (1992c), "Proposal for a council directive introducing a tax on carbon dioxide emissions and energy", *COM* (92) 226 final.

EC (1992d), "Proposal for a council directive to limit carbon dioxide emissions by improving energy efficiency (SAVE programme)", *COM* (92) 182 final.

EC (1992e), "Council directive 92/75/EEC on the indication by labelling and standard product information of the consumption of energy and other resources by household appliances", *OJL* 297 13.10.92, pp.16-19.

EC (1993), "Council directive 93/76/EEC to limit carbon dioxide emissions by improving energy efficiency", *OJL* 237 22.9.93, pp.28-30.

ECSC-EEC-EAEC (1987), *Treaties establishing the European Communities*, Office for Official Publication of the European Communities, Luxembourg.

Hennicke, P., Johnson, J., Kohler, S., Seifried, D. (1987), *Die Energiewende ist möglich*, Öko-Institut, Freiburg.

IEA (1992), *Climate change initiatives*, OECD/IEA, Paris.

Verhoeve, B., Bennett, G. and Wilkinson, D. (1992), *Maastricht and the environment*, Institute for European Environmental Policy, Arnhem and London.

WCED (1987), *Our Common Future*, Oxford University Press, Oxford.

Printed and bound by CPI Group (UK) Ltd, Croydon, CR0 4YY

21/10/2024

01777088-0005